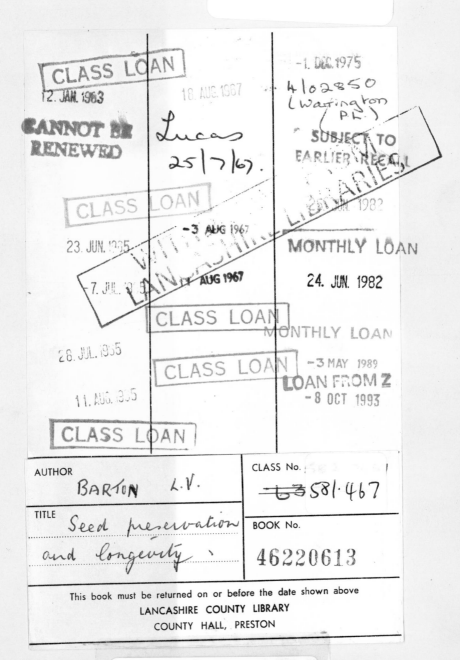

PLANT SCIENCE MONOGRAPHS

Edited by

Professor Nicholas Polunin

––––––––––

SEED PRESERVATION AND LONGEVITY

PLANT SCIENCE MONOGRAPHS

Uniform with this volume will be:

Biology of Mycorrhiza	J. L. Harley
Encyclopedia of Weeds and their Control	L. J. King
Grassland Improvement	A. T. Semple
Mangroves of the World	V. J. Chapman
Microbiology of the Atmosphere	P. H. Gregory
Mutations and Crop Improvement	Å. Gustafsson
**Plant Growth Substances*	L. J. Audus
Plant Life and Nitrogen	G. Bond
**Salt Marshes and Salt Deserts of the World* . .	V. J. Chapman
Sex in the Lower Organisms	H. P. Papazian

FURTHER TITLES ARE UNDER CONSIDERATION

THE feeding and clothing of the world's teeming millions can continue to keep abreast of population increases through the help of effective application of research in the plant sciences. The publication of this research, by which means a scientist or technologist makes his findings known to workers elsewhere, tends to be scattered in literally hundreds of botanical and agricultural journals emanating from most of the countries of the world. Often it appears in such polyglot arrays of fragments that it is extremely difficult to bring together even in some narrow 'line' of endeavour. Consequently advances are slowed and interests unnecessarily divided, scientific and human progress being thereby retarded.

The present series of 'monographs' is designed to remedy these deficiencies in especially important or attractive specialities, by publishing individual book-length accounts of the entire background and current progress in their fields. Such detailed surveys, being fully documented and plentifully illustrated, should prove of real value to the world at large in constituting the bases for further advances on the ever-expanding horizons of scientific research, and so lead to improved productivity and, ultimately, standards of living. They are prepared by specialists usually of international reputation for their work in the field chosen, and often culminate a lifetime of active investigation. Being as up-to-date as possible, they will often embody significant advances not previously published.

* Already published and available.

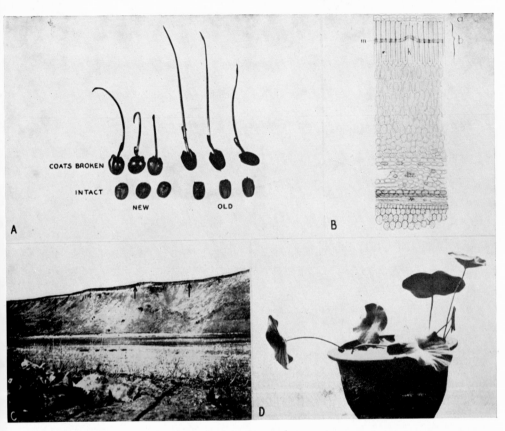

PLATE I

A, Relative rate of growth of seedlings produced by freshly harvested and old seeds of the Sacred or Indian Lotus (*Nelumbo nucifera*).

B, Structure of seed-coat of *Nelumbo* showing water-resistant layers, namely, (*a*) epidermis and (*b*) part of palisade layer to the 'light-line' (*m*), all greatly magnified.

C, 30- to 60-foot bank left by the cut of the river; seeds were located in strata 3 to 6 feet from the top of the precipice, as indicated by arrows.

D, Plant grown from one of the old seeds.

PLANT SCIENCE MONOGRAPHS
edited by
Professor Nicholas Polunin

SEED PRESERVATION AND LONGEVITY

By

LELA V. BARTON

M.A., Ph.D. (Columbia)
Boyce Thompson Institute for Plant Research,
Yonkers, New York, U.S.A.

1961
LONDON
LEONARD HILL [BOOKS] LIMITED

INTERSCIENCE PUBLISHERS, INC.
NEW YORK

LONDON
Leonard Hill [Books] Ltd.
9 Eden Street, N.W.1

NEW YORK
Interscience Publishers Inc.
250 Fifth Avenue, New York 1

FIRST PUBLISHED IN GREAT BRITAIN 1961

PRINTED IN GREAT BRITAIN AT
THE UNIVERSITY PRESS
ABERDEEN

INTRODUCTION

ONE of the most important factors contributing to maximum agricultural crop production per unit of land is the use of good seeds.* Such seeds should have a high purity value, i.e. they should be free from all other seed types, whether of weeds or of cultivated plants, and their germination capacity and vigour should be high, thus providing the best possible guarantee of a good seedling stand—in spite of diseases, insects, competition of weeds, and unfavourable weather conditions. This fact has, of course, been recognized for a long time; but the systematic storage of seeds to maintain viability, and the routine testing and analysis of seeds for planting value, dates from only about the beginning of this century.

Increased interest in, and knowledge of, the seed problem have resulted in the establishment of private, state, and federal seed-testing laboratories and the formation of national and international rules for seed testing. Along with improved methods of testing has come the necessity for proper storage facilities to maintain viability. It is important to seedsmen to be able to keep surplus supplies for sale in later years. In some instances this has been done without the benefit of proper storage conditions for maintaining high-quality seeds. As a result, everyone who buys seeds is apt to get lots which are of low quality or which fail to germinate. As more information is secured on the storage requirements, reputable seedsmen are building dehumidified or cold-storage rooms. The possibility of holding seeds in storage for at least two years reduces the amount of land needed for seed production, as crops can then be alternated.

Many coniferous trees do not set seed every year: indeed there may be an interval of as much as ten years between crops. This poses a serious problem in reafforestation projects which depend on a constant supply of seeds for nursery use. Also, direct reseeding of forest land could be seriously hampered by the lack of a seed supply in any given year. With the benefit of below-freezing storage, it is safe to hold conifer seeds for periods adequate to cover these needs. Such low temperatures automatically control insect damage, which is considerable in many seeds. Insect control has usually depended on the presence of certain insecticides which, however, introduce the further problem of injury to the seeds or to the persons handling the seeds.

The possibility of long-term storage is a boon to persons concerned

---

* It has not been found necessary to distinguish in this work between seeds and seed-like fruits (Ed.).

with genetics or with plant introduction, for it provides a constant source of valuable seed stock without the necessity of growing large numbers of plants and controlling seed production—a laborious and difficult process.

Granted that good storage conditions for a certain seed have been determined, there are still other questions which arise. Will the seeds from storage, especially low-temperature storage, survive removal for packeting, which must be done well in advance of the time the consumer buys and plants them? Furthermore, after packeting, seeds are sometimes stored under very unfavourable conditions on the retailers' shelves. Some experimental results on this problem have demonstrated the efficacy of moisture-proof packets. The best storage condition, determined for a specific seed, provides it with maximum resistance to harmful conditions following packeting for sale. Additional work needs to be done in this field, using different kinds of seeds; but the principles already determined will probably apply to many species.

The quality of plants which can be expected from old seeds is another matter of importance. An increasing body of evidence shows that the performance of any seed depends less upon its age than upon the conditions under which it has been stored.

The absolute dependence of the world's population on plants for food, and the realization that most plants are grown from seeds, should be sufficient to dramatize the importance of conserving our seed supply. A basic bibliography of literature on the subject is to be found in several general articles or books (Franck, 1928; Crocker, 1938; Barton & Crocker, 1948; Crocker, 1948; Crocker & Barton, 1953; and Owen, 1956).

PREFACE

Spermatophytes,* or seed plants, are more important to man than any other living plants, for they are best adapted to cultivation and furnish, directly or indirectly, most of the food and clothing for the entire world. These plants are distinguished from lower groups by the production of seeds. By the development and establishment of the seed habit, the reproduction of the individual plant and the continuance of the species are assured. Furthermore, the seed is an excellent method for multiplication and distribution of the species. The seed plants are divided into *Gymnosperms* and *Angiosperms*. The former produce 'naked' seeds, which are so called because they are borne on the surface of a scale, usually of a cone, as in pine (*see* Figure 1, A¹). Gymnosperms are woody perennial forms comprising a group of about 500 species, most of which are evergreens. They include many commercially important types that are valuable for timber and resins of various kinds.

In the Angiosperms, the seeds are borne in an enclosed structure or fruit developed from the ovary. This fruit may be dry or fleshy, and may contain one or several seeds. The Angiosperms dominate the modern plant world, being represented by some 250,000 species. They are divided into two groups, *Monocotyledons* and *Dicotyledons*. There are several structural differences between these two groups; but, from the point of view of seed distinctions, the former possess embryos with only one terminal cotyledon, while the embryos of the latter have two lateral cotyledons. All grasses, including the commercially important cereal and many forage crops, are monocotyledonous, while most of the vegetables, flowers, and deciduous trees belong to the Dicotyledons.

Seeds are usually produced as a result of pollination and fertilization. Flowering is the first step in such seed production. In the ordinary way, pollen, produced in great abundance in the anthers of the stamens—the male parts of the flower—is carried by wind, insects or other means to the stigma, which is the tip of the pistil or female part of the flower. Here the pollen is commonly held by a special secretion in which the pollen grain germinates to produce a tube which carries the male nucleus down the style of the pistil to the ovary, which contains the ovules. The pollen tube penetrates the ovule and normally its germ nucleus fuses with the egg nucleus of the ovule to form the embryo of the seed.

Embryos vary greatly in size and appearance, but, with few exceptions, they are made up of the same organs: a plumule or rudimentary shoot,

* Technical terms, etc. are explained in the Glossary commencing on p. 162.

vii

cotyledons or seed-leaves, hypocotyl between the cotyledons and plumule, and the radicle or rudimentary root. The plumule, hypocotyl, and radicle form the axis of the embryo. In Dicotyledons the cotyledons may remain underground upon germination of the seed, or they may be pushed above ground by the growth of the hypocotyl to form fleshy green leaves as in bean. The single cotyledon (scutellum) of grasses (Monocotyledons) remains as a food-absorbing organ within the germinating grain. In some seeds, as for example those of orchids or holly, the embryo is an undifferentiated mass of cells at the time the seed matures in the plant. Such embryos must develop the structures described above before they are ready to form a new plant. This results in delayed germination when the seed is planted.

The endosperm, or food stored for use by the young embryo, also commonly develops in the seed following the conjugation of a second male nucleus from the pollen tube with two other nuclei in the ovule. This means a double fertilization, one resulting in the young plant or embryo and the other in the development of food for the young plant. Such double fertilization occurs in most Angiosperms, and results in the formation of the endosperm, a definite structure containing stored food. However, the endosperm is not always present in the mature seed, as it may have been used up during maturation of the seed. In the bean seed, for example, most of the stored food has been absorbed by the cotyledons which grow to fill most of the seed cavity. Maize, wheat and other cereals, on the other hand, have abundant endosperm.

Just as the seed is the ripened ovule composed of embryo and seed-coats together with stored food, so the fruit in which the seed is borne is the ripened ovary together with other parts of the flower which may have become associated with it. There are many different kinds of fruits, but, for the present purpose, it is perhaps sufficient to classify them as simple, aggregate, accessory, or multiple.

Simple fruits, or those resulting from the ripening of a single ovary, may be further classified into dry or fleshy. Dry fruits are dehiscent or indehiscent. Many-seeded dehiscent fruits include pods and capsules of various kinds which open when dry to liberate the seeds within them. Pods may open through one suture (larkspur, columbine, milkweed); one or two sutures (legumes, *see* Figure I D); a circular line cutting off the upper part of the fruit as a lid (plantain); or separation of two valves of a capsule (mustards). Indehiscent dry fruits are mostly one-seeded or have very few seeds. These fruits are commonly known as 'seeds' because they do not open to discharge the true seeds, but adhere tightly to them. One of the most common of these dry, indehiscent fruits is the grain, or caryopsis. Here the single seed completely fills the fruit, as in maize (*see* Figure I E) and wheat, for example. In addition to the fruit-coats, grass 'seeds' often have small bracts (the lemma and palea) of the original floret associated with them. Other indehiscent dry fruits are samaras or keys, which are provided with a wing for dispersal. Examples of this type are the fruits of ash, with a terminal wing, or of elm, where the wing forms a margin around the entire fruit, or of maple (*see* Figure, 1 c), which is a double samara or pair of fruits conspicuously winged from the apex. Achenes are small fruits, which often have appendages (dandelion, lettuce) to act as agents of dispersal. The actual fruits of the strawberry are also achenes—in this case produced on a fleshy flower-receptacle which really has no connection with the pistil or female part of the flower. A nut is another hard, one-celled and one-seeded, indehiscent fruit (beech, chestnut, oak).

Fleshy fruits which are also indehiscent may be fleshy throughout, with or without a firm rind or shell, or they may be fleshy externally and hard or stony internally. Among the latter type are drupes or stone fruits such as those of cherry or peach. The pomes (apple, pear, quince) are fleshy fruits composed of two or several carpels of bony or cartilaginous texture enclosed in flesh which morphologically belongs to the calyx and receptacle of the flower from which the fruit was produced (*see* Figure 1, B, B^1). The fruit of the gourd and squash is a pepo, a fleshy fruit with a hard rind. Other examples of simple fleshy fruits are the hesperidium, a special type of berry with a leathery rind represented by fruits of orange, lemon and lime, and the berry itself which includes the fruits of grape, currant, banana and tomato, in which the fruit is fleshy throughout.

Aggregate fruits are those in which a cluster of pistils, all from one flower, is crowded on the flower receptacle in one mass, as in raspberry and blackberry. Multiple or collective fruits, on the other hand, are those which result from the aggregation of several flowers into one mass (pineapple, mulberry, fig).

The seed industry, one of the most important in the world because it is at the heart of the survival of man, involves varied investments and skills. It requires the services of geneticists and plant breeders to develop and select varieties superior in performance and in yield of vital food and clothing materials. Plant physiologists are constantly working to determine the effect of nutrition and environmental conditions and many other factors on growth, seed-set and quality. Plant pathologists aim to control the diseases of plants, and entomologists study the insects which affect plant life in a never-ending battle to reduce the ravages of these pests. Growers must practise the highly specialized techniques of seed production. Special attention must be paid to the harvest and curing of seeds. Germination problems are many, and they must be determined for each seed type. Testing and analyses of seed-lots have become problems of major concern to every country, and most countries now have set certain standards which must be met before seed can be offered for sale in them. Seed marketing forms an industry in itself. Finally, equipment manufacturers must provide the apparatus and materials required for the functioning of this complex industry.

Some idea of the extent of the seed industry in the world may be obtained from a few statistics of seed production in the United States of America. In 1956, 75,950,000 acres of maize (U.S. corn) were harvested to yield 3,451,292,000 bushels of grain; wheat harvested from 49,817,000 acres produced 997,207,000 bushels of seed. In 1955, 212,390,000 lb. of clean alfalfa seeds valued at $43,752,340 were grown on 1,392,500 acres. During the same year 204,757 acres were planted in vegetables which yielded 216,294,000 lb. of seeds. The bean crop alone in 1955 required 1,502,000 acres to produce 1,664,900,000 lb. with a market value of $116,709,490.

In addition to the seeds grown for the production of food, clothing, and shelter, for man and domesticated animals, there is a large and expanding flower-seed industry. Tree seeds also add their importance to those already described. Tons of such seeds are needed each year for reafforestation of lands throughout the world.

Members of the seed industry have the grave responsibility of furnishing a good start for growth of important crops by providing seeds of desirable strains and high quality, year after year. It is always necessary to hold seeds from the time of harvest until planting time, and it is often desirable, in the interest of economy, to store excess harvest for one or several seasons. It becomes necessary then to know safe storage conditions, which vary from seed type to seed type. Also, safe shipment of fresh seeds is often a problem—especially if they must be routed through hot, humid climates. Over the years, seed physiologists and other individuals concerned have tried to uncover some of the secrets held within the coats of seeds, both viable and dead, in order to determine their keeping qualities under various conditions.

This book is an attempt to combine in one volume the available information concerning the longevity of seeds under natural and controlled conditions. The scientific aspects of the problem are considered in so far as they are known, and the practical applications of the existing knowledge are pointed out. No attempt has been made to list all references to published articles on the general subject of 'Seed Preservation and Longevity', for this would be tedious and repetitious. Rather, literature has been selected which, in the opinion of the author, is representative of the various phases that should be covered in such a work.

The author is grateful to Dr. George L. McNew, Managing Director of the Boyce Thompson Institute for Plant Research, Inc., for his interest and help; to Mrs. Jean G. Fine, Miss Joyce Fitzgeorge, and Dr. Norma E. Pfeiffer, for the checking of literature and for proof reading; and to Messrs. L. P. Flory and William G. Smith, Jr., for photographing and arranging plates. The author is also grateful to the following for permission to reproduce certain figures from original research articles: Drs. G. C. Papavizas and C. M. Christensen, Fig. 7 and Plate 2; Drs. N. Armolik, J. G. Dickson, and A. D. Dickson, Plate 3; Associated Seed Growers, Inc., Plates 6 and 7; Dr. G. Lakon, Figs. 12 and 13; Dr. M. Simak, Plate 9;

Drs. E. H. Toole, V. K. Toole, and H. A. Borthwick, Plate 10*a*, *b*; and Drs. H. Gunthardt, L. Smith, M. E. Haferkamp, and R. A. Nilan, Plate 13. An acknowledgment is given in the legend for each of these figures. Acknowledgments for permission to reproduce tables from various publications are also given in the text.

LELA V. BARTON

BOYCE THOMPSON INSTITUTE FOR PLANT RESEARCH, INC., YONKERS, NEW YORK

CONTENTS

LIST OF TABLES

LIST OF ILLUSTRATIONS
PLATES

FIGURES IN TEXT

FIG. I.—Some representative seeds and fruits. A, A' Cone and winged seed of Sugar Pine (*Pinus lambertiana*). B, B', Pome and seeds of Pear (*Pyrus communis* cultivar William Bartlett). C, Leaf and double samara containing seeds of Sugar Maple (*Acer saccharum*). D, Pods and seeds of Garden Pea (*Pisum sativum*). E, Grains of Maize (*Zea mays*), the one on the left germinating.

RECORDS OF OLD SEEDS

THE life-span of seeds has been a matter of interest, conjecture, and fantastic report, for more than a hundred years. Turner (1933), in his review of literature on the subject, says that Theophrastus mentions that 'of seeds some have more vitality than others as to keeping' and 'in Cappadocia at a place called Petra, they say that seed remains even for forty years fertile and fit for sowing'. All of the early reports, including those made in the first part of the present century, were concerned with seeds of more or less established ages, which were found, for example, on herbarium specimens or in old cupboards. Very little if anything was known about the origin of these seeds and the specific storage conditions which may have affected their longevity. Furthermore, methods used for germination tests may not have been, and probably were not, those which were best for measuring viability. In spite of these handicaps, some valuable data were obtained.

Ewart (1908), in his treatise *On the Longevity of Seeds*, pointed out that probably few sections of human knowledge contained a greater number of contradictory, incorrect, and misleading observations than prevail in the works dealing with seed viability, and that knowledge of the subject was still in an incomplete and indeed fragmentary condition at that time. On the basis of reports of others and some work of his own, Ewart divided seeds into three biological classes according, as he said, to their duration of life under favourable conditions. These were short-lived or 'microbiotic' seeds, with a life-span not exceeding three years, 'meso-biotic' seeds which may live from three to fifteen years, and 'macrobiotic' seeds which may retain viability for fifteen to over a hundred years. Under the macrobiotic classification he listed 137 species in 47 genera of the family Leguminosae; 15 species in 8 genera of Malvaceae; 14 species in 4 genera of Myrtaceae; 3 species in 2 genera of Nymphaeaceae; 3 species in 3 genera of Labiatae; 2 species in 2 genera of Irideae; and 1 species in each of the families Euphorbiaceae, Geraniaceae, Goodeniaceae, Polygon-aceae, Sterculiaceae, and Tiliaceae.

One of the earliest records of germination of old seeds is that of Candolle (1846). Seeds of 368 species representing various families were collected, principally from the botanical garden of Florence in 1831, and were tested in 1846 when nearly fifteen years old. They had been kept in a cabinet protected from high humidity and extremes of temperature. Only seventeen of these species germinated when the seeds were planted in soil, and only one of these, *Dolichos unguiculatus*,* showed more than 50 per cent germination. Leguminosae (5 out of 10 species tried) and Malvaceae (9 out of 45 species tried) accounted for 14 of the 17 viable lots of seeds. The seeds tested represented 180 annuals, 28 biennials, 105 perennials, and 44 woody plants, together with 11 unidentified seed-lots. The author concluded that the figures obtained seemed to prove that woody species preserve their viability longer than others, and that biennials, none of which were found to be viable, deteriorated most rapidly.

TABLE I

RECORD OF OLD SEEDS

(from Becquerel, 1934, p. 1662)

Macrobiotic species	Date collected	Seeds growing in 1906		Seeds growing in 1934		Determined longevity (yr.)	Probable longevity (yr.)
Mimosa glomerata Forsk.	1853	5 out of 10		5 out of 10		81	221
Melilotus lutea Gueld.	1851	3	„ 10	0	„ 10	55	—
Astragalus massiliensis Lam.	1848	0	„ 10	1	„ 10	86	100
Cytisus austriacus L.	1843	1	„ 10	0	„ 10	63	—
Lavatera pseudo-olbia Desf.	1842	2	„ 10	0	„ 10	64	—
Dioclea pauciflora Rusby	1841	1	„ 10	2	„ 10	93	121
Ervum lens L.	1841	1	„ 10	0	„ 10	65	—
Trifolium arvense L.	1838	2	„ 10	0	„ 10	68	—
Leucaena leucocephala L.	1835	2	„ 10	3	„ 10	99	155
Stachys nepetifolia Desf.	1829	1	„ 10	0	„ 10	77	—
Cytisus biflorus L'Hérit.	1822	2	„ 10	0	„ 10	84	—
Cassia bicapsularis L.	1819	3	„ 10	4	„ 10	115	199
Cassia multijuga Rich.	1776	—		2	„ 2	158	—

Becquerel (1907) tested seeds of 500 species from the seed collection of the Museum of Natural History in Paris. The authenticated ages of these seeds were known to be between 25 and 136 years. Germinations were obtained from four families: Leguminosae, Nymphaeaceae, Malvaceae, and Labiatae. Twenty of these seeds were from twenty-eight to eighty-seven years old. Later, Becquerel (1934) made another test on the viability of the old collections of seeds which he had used in 1906, with the results shown in Table I. All of these seeds which germinated were of Leguminosae, except for *Lavatera* (Malvaceae) and *Stachys* (Labiatae). The seeds of *Cassia multijuga* germinated after 158 years of storage—apparently

---

* Authorities for scientific plant names, which may be needed to indicate the precise sense in which the latter are employed, are usually given only when original observations are concerned, or where specified by the authors of cited works.—Ed.

2

a record at that time. Turner (1933) found that seeds of *Anthyllis vulneraria* and *Trifolium striatum* remained viable for ninety years and those of *Cytisus scoparius*, *Melilotus alba*, *Lotus uliginosus*, and *Trifolium pratense* for eighty-one years.

Schjelderup-Ebbe (1936) had access to two collections of old seeds in the Botanical Museum of the University of Oslo. The larger and older collection was that of N. G. Moe, Head Gardener of the Botanical Garden from 1857 to 1892. It contained seeds collected mainly during the years 1820 to 1892 as well as some older seeds. The second collection was that of A. G. Blytt, collected mainly during the period of his professorship, 1880 to 1898. Seeds of the first collection were stored in strong paper bags with the name of the species and date of collection written on the outside. The second collection was stored in corked bottles with the species, time, and place of collection written on labels pasted on the bottles. The germination tests were run during the years 1932 and 1933, so the ages of the seeds tested ranged from approximately 34 to 112 years. The seeds included Gymnosperms and Angiosperms; the latter group included monocotyledons as well as dicotyledons. In all, 1,254 different batches of seeds were tested and relatively few of these included duplication of species. Out of these, seeds of only fifty-three species showed any germination. The oldest living seeds found in these studies were those of *Astragalus utriger*, eighty-two years old, which gave 6 per cent germination. The author stated that he had added two families of plants to those bearing macrobiotic seeds: Cannaceae, with *Canna paniculata* seeds sixty-nine years old; and Thymelaeaceae, with *Daphne mezereum* seeds thirty-five years old. Seed size appeared to be unimportant, as both very small and very large seeds had lost their viability. The only report of an older viable seed from stored collections was from those of the British Museum (Anon., 1942), which recorded the germination of a seed of *Nelumbo* after about 250 years * of storage as well as of seeds of *Albizzia julibrissin* after 149 years.

Exell (1931) reported that Robert Brown found 150-year-old viable seeds of *Nelumbo nucifera* (syn. *Nelumbium speciosum*), the Sacred or

---

* It has, unfortunately, proved impossible to determine the exact age of these *Nelumbo* 'seeds' in the British Museum, where Mr. A. W. Exell now points out that *N. lutea* may have been the species involved. The seeds were from the Sloane collection, and concerning them J. E. Dandy (*The Sloane Herbarium*, British Museum, London, 1958, p. 18) remarks, 'Among the most interesting objects in this collection were the seeds of the North American *Nelumbo lutea* (Willd.) Pers. (No. 8517), which were germinated by Robert Brown in 1848, having been sent to Sloane about 1727. This is a remarkable instance of seeds' having germinated at a considerable period since they were collected. Other specimens of *Nelumbo* which, like the preceding, were in the Department of Botany, were germinated by Brown between 1843 and 1855; these may have represented the same No. or possibly No. 8110, acquired by Sloane from Petiver, and hence collected before 1718, the date of Petiver's death. No. 8517 is no longer represented in the collection, but specimens of No. 8110 still remain.' It is to be hoped that some at least of these last can be induced to germinate after 1968, so confirming the full quarter-millennium for seeds in dry storage.—Ed.

3

Indian Lotus, in 1843–55. In 1925, I. Ohga visited London and made further tests with these same collections of seeds. They had lost their power to germinate some time between 1855 and 1926. Ohga, however, succeeded in germinating seeds of the same species that had been collected in a layer of peat in a naturally drained lake-bed in Manchuria (Ohga, 1923, 1926), *see* Plate 1(*c*). From the information he was able to obtain about the site in which these old seeds were found, Ohga thought them to be 'at least four hundred years old' (Exell, 1931). Libby (1951), using some of the original lot of Ohga seeds which were still viable at that time, tested them with the radiocarbon technique and estimated that they were 1,040 ± 210 years of age. These are the oldest known viable seeds, *see* Plate 1(*d*). Libby reported, later (1954), that Ohga found three viable Lotus seeds associated with the remains of a canoe discovered about twenty feet below the surface of a lake in Kemigawa, Japan, in 1951. The average age of the wood found in this canoe was 3,075 ± 180 years, but this of course does not necessarily indicate the age of the seeds. There is, however, no doubt that the seed of *Nelumbo nucifera* is capable of germinating after long periods of storage in a natural condition.

It should be noted that the old Lotus seeds did not germinate without some treatment to make the hard coat permeable to water. This was done by concentrated sulphuric acid treatment, or by the simple expedient of filing the coat. After such coat treatment the relative growth of the seed-lings from the old seeds was greater than that from fresh seeds of the same species (Ohga, 1926*b*) as shown in Plate 1(*a*). Apparently normal plants from these old seeds have been grown in many parts of the world (cf. preceding page and Plate 1(*d*)). Reports claiming that they are 50,000 years old, however, are apparently not valid (Anon., 1951).

The records given above of longevity of old seeds in herbaria and cupboards include seeds with impermeable coats which are especially characteristic of the family Leguminosae. Because the coats must be made permeable to water before germination can proceed, it is very likely that some of the tests were invalid, for not all of the workers appeared to have been aware of this necessity. Much later work under controlled storage conditions has pointed to the efficacy of the sealing process to maintain the moisture content favourable for retaining viability. It is as though each seed of great natural longevity had been individually sealed, so that the species could be ensured of distribution in time. The importance of coat characters in seed longevity has been recognized from the earliest reports and was emphasized by Crocker (1909) as early as 1909.

EGYPTIAN MUMMY WHEAT

The impermeable coat characteristic of long-lived seeds would seem to discredit the perennial reports of the viability of seeds from Egyptian tombs. For the most part these claims have been made for wheat seeds,

4

though pea seeds of this ancient vintage have also come in for a share of attention. An article in the *Gardeners' Chronicle* under the date of Saturday, November 11, 1843 (Anon., 1843), reported the germination of one of twelve wheat seeds taken from an alabaster vase in a tomb not visited by man for 3,000 years. The resulting plant grew to maturity and produced twenty-seven grains which, in turn, germinated and produced vigorous plants. M. F. Tupper, who germinated the seed, remarked in conclusion (Anon., 1843, p. 787), 'If, and I see no reason to disbelieve it, this plant of wheat be indeed the product of a grain preserved since the time of the Pharaohs, we moderns may, within a little year, eat bread made of Corn which Joseph might have reasonably thought to store in his granaries, and almost literally snatch a meal from the kneading-troughs of departing Israel.'

By far the majority of reporters on seeds from tombs, however, take the view that the seeds could not possibly be viable or that a hoax of some kind is involved (Nicholson, 1932; Anon., 1933; Cifferi, 1942; Bunker, 1946). A description of a 'mummy' pea, so-called for no other reason than that someone years before had claimed that the original pea of this variety had been found in an Egyptian mummy, has been given by O. A. White (1946), who discredits the story. Examinations of the actual seeds removed from tombs have also been made. Luthra (1936) found some of the wheat grains completely carbonized while still retaining their original shape. When moistened, these grains changed to a powdery black ash. Barton-Wright *et al.* (1944) found that barley from King Tutankhamen's tomb was extensively carbonized, but the germ with its scutellum and embryo components was still intact. However, the old grain had apparently lost a considerable amount of weight and density as compared to that of fresh English barley. Also, microbiological assay revealed that riboflavin and nicotinic acid were still present in the old barley, together with an increased acidity. The authors thought it was possible that preservation of the barley was helped by the oxygen-free atmosphere created by its own respiration and the uptake of oxygen by other products stored with it. Nothing was said about the germination of these seeds. Åberg (1950) described the taxonomic characters of samples of barley and wheat about 5,000 years old, taken from the Saqqara Pyramid. An article in *Nature* (Anon., 1934) reviews some of the reports on the viability of such seeds and says that popular belief in the viability of wheat grains which have been interred in ancient tombs, sometimes thousands of years old, has been severely shaken by morphological and physiological tests on genuine mummy wheat, and also by bringing into question the authenticity of other so-called specimens.

Radiocarbon dating has revealed that some wheat and barley grain from Egypt is 6,391 ± 180 years old (Libby, 1951), but no claim is made for the viability of such seeds.

In spite of all the evidence to the contrary, people will continue to be interested in, and some of them will continue to believe in, the viability of seeds from ancient tombs, though it is likely that any such viable seeds were carried in with modern packing or deliberately placed there as a hoax.

Much more will be said in the later pages of this book about the life-span of seeds with permeable coats. It will be seen that it is possible to extend the life of the seed by providing it with the proper storage conditions. However, we have no evidence at present that wheat seed will remain viable for more than twenty-five to thirty years.

For the most part, early records of aged seeds were obtained from herbaria or storage cupboards. These were from seeds which just happened to be available for testing. One of the earliest experimental designs to test the effect of various storage conditions on seed viability was that described by van Tieghem & Bonnier (1882). They stored some leguminous, castor oil and flax seeds in tubes in air or carbon dioxide, as well as in open tubes—all at room temperature. After two years, the seeds stored in open containers had decreased little or much in germination, depending on the species, while those sealed in air or carbon dioxide germinated less readily or had lost viability completely. In the light of future, more accurately controlled tests, these data mean little because of the lack of water-content measurements. All of the seeds exposed to air had gained weight at the end of two years, while those in sealed containers had not. This perhaps indicated a high moisture content at the beginning of the storage period—a condition which would bring about rapid deterioration in sealed storage.

Kondo (1926) recognized the importance of variety and of climatic conditions on the keeping quality of seeds. He stored seeds of Japanese cultivated plants in sacks in the laboratory, and found their life duration very short—much shorter than was recorded for seeds of Australia, Europe, and America. He believed this to be due to the moist, hot summer climate of Japan.

In order to investigate the effect of climatic conditions, Duvel (1904) stored seeds of some common cultivated plants in laboratories in Puerto Rico, and in different states in the United States of America (Florida, Alabama, Louisiana, Indiana, New Hampshire, and Michigan). From these studies he concluded that precipitation is a factor of much greater importance than temperature. Immature seeds and seeds harvested in wet weather did not retain viability as long as mature seeds and seeds ripened in dry weather. The life-span varied with the family, genus, and species. Under good storage conditions the life-span could be lengthened, but never, in his opinion, for centuries.

6

LIFE-SPAN OF SEEDS BURIED IN SOIL

It is, perhaps, not surprising that seeds which are dry and, as a consequence, have their metabolic processes greatly reduced, are still capable of germination after years of storage. The extended life-span of some species in the soil, however, is more difficult to explain. There are many reports of seeds germinating after ten or more years in the soil. These reports have been based, for the most part, on the appearance of plants not common to the vicinity—usually on newly excavated or ploughed soil, or on bombed sites following a war. Numerous investigators have been concerned with the viability of seeds buried in the soil for long periods of time. Weed seeds, especially, have been the subject of many studies, because of the difficulty of eradicating all of the plants from cultivated gardens. Turrill (1957) stated that it has been proved at Rothamsted that soil under pastures contained buried and viable seeds of arable weeds after periods of as much as 300 years in one area and of thirty to forty years in others.

While much of the information about the viability of buried seeds has not permitted the actual determination of the age of the seeds, some controlled experiments have been conducted. Perhaps the earliest of these was one started in 1879 by Beal (Darlington, 1951). Seeds of twenty different wild species were mixed with sand and buried in uncorked pint bottles, the mouths of which were tilted downwards to prevent filling with water. Burial was in sandy soil approximately 18 in. below the surface. A summary of the results obtained after five to seventy years is shown in Table II. It will be noted that some of the seeds did not withstand burial for as much as five years. Again, seeds of eleven species were still alive after twenty years. After forty years' burial, seeds of eight species, namely *Amaranthus retroflexus, Ambrosia elatior, Brassica nigra, Lepidium virginicum, Oenothera biennis, Plantago major, Portulaca oleracea,* and *Rumex crispus* all produced seedlings, while after seventy years only *Oenothera biennis* and *Rumex crispus* were capable of germination. Testing was done by transferring the contents of the storage bottle to a 'flat' box of sterilized soil, which was placed in the greenhouse. Seedlings were identified after they had grown sufficiently to make this possible. It will be noted that seedlings of all viable species were not obtained at each testing. Also, *Verbascum blattaria* and *Silene noctiflora*, which were not included in the original storage, had appeared after fifty or sixty years. The table gives a record of all the tests

7

TABLE II

RESULTS OF ALL TESTS TO DATE IN BEAL'S BURIED SEED EXPERIMENT

(from Darlington, 1951)

Name of species tested	5th yr. 1884	10th yr. 1889	15th yr. 1894	20th yr. 1899	25th yr. 1904	30th yr. 1909	35th yr. 1914	40th yr. 1920	50th yr. 1930	60th yr. 1940	70th yr. 1950
Amaranthus retroflexus	+	+	+	+	+	+	o	+	o	o	o
Ambrosia elatior	o	o	o	o	o	o	o	+	o	o	o
Brassica nigra	o	+	+	+	+	+	+	+	+	o	o
Bromus secalinus	o	o	o	o	o	o	o	o	o	o	o
Bursa bursa-pastoris	+	o	+	+	+	+	+	o	o	o	o
Erechtites hieracifolia	o	o	o	o	o	o	o	o	o	o	o
Chamaesyce maculata	o	o	+	o	o	o	o	o	o	o	o
Lepidium virginicum	+	+	o	+	+	+	+	+	o	o	o
Agrostemma githago	o	o	+	o	o	o	o	o	o	o	o
Anthemis cotula	+	+	o	o	+	o	~	o	o	o	o
Malva rotundifolia	+	o	+	+	o	o	o	o	o	o	o
Oenothera biennis	+	+	+	+	+	+	o	+	+	o	o
Plantago major	o	o	+	o	o	o	o	+	o	o	o
Polygonum hydropiper	o	+	+	+	+	+	o	o	o	+	+
Portulaca oleracea	o	o	+	+	+	~	o	+	+	o	o
Rumex crispus	+	+	+	o	+	o	o	+	o	o	+
Chaetochloa lutescens	+	+	+	+	+	+	o	o	+	o	o
Alsine media	+	+	+	+	+	+	o	o	o	o	o
Trifolium repens	o	~	o	o	+	+	o	o	o	+	o
Verbascum thapsus	+	+	+	+	o	o	o	o	o	o	c
(Verbascum blattaria)		+	o	o	o	o	o	o	o	+	+
(Silene noctiflora)		~	+	+	o	o	+	o	+	+	+

8

up to 1950 inclusive, and indicates the number of times each species has produced at least one seedling.

A similar experiment, but one which was terminated after thirty-nine years, was a project of the Seed Testing Laboratory of the United States Department of Agriculture, and was started by Duvel in 1902. The final results of this experiment have now been given by E. H. Toole & Brown (1946). Seeds of 107 species representing both wild and cultivated plants were buried in thirty-two sets in sterile soil in flower pots with porous clay covers at depths of 8, 22, and 42 in. Germination tests have been reported after burial for 1, 3, 6, 10, 16, 20, and 39 years. Of the 107 species buried in 1902, 71 germinated after one year, 61 after three years, 68 after six years, 68 after ten years, 51 after sixteen years, 51 after twenty years, 44 after thirty years, and 36 after thirty-nine years. There was a general tendency to lower germination of seeds from the 8-in. depth than from the 22-in. depth, and to the highest germination from the 42-in. depth. The sixteen species, representing ten plant families, having the highest germination after thirty-nine years (more than 15 per cent from at least one depth) were: *Abutilon theophrasti, Ambrosia artemisiifolia, Convolvulus sepium, Datura stramonium, Ipomoea lacunosa, Lespedeza intermedia, Nicotiana tabacum, Oenothera biennis, Onopordum acanthium, Phytolacca americana, Potentilla norvegica, Robinia pseudoacacia, Rudbeckia hirta, Solanum nigrum, Trifolium pratense,* and *Verbascum thapsus.* Of the twenty other species that showed some life after thirty-nine years, eighteen exhibited not more than 6 per cent germination from any storage depth. Most of the seeds of cultivated plants died after one year in the soil, but *Nicotiana tabacum* and *Trifolium pratense* survived for thirty-nine years. Seeds of wild plants, on the other hand, were more resistant to burial than were those of cultivated plants.

A buried seed experiment set up to last for forty-six years was started in 1932 by Goss, using twelve weeds of California (Goss, 1939). Very low germination obtained in 1933 and 1934 was attributed to failure to supply the special germination requirements of the seeds, including treatment of impermeable coats. Many of the seeds were dead by 1938. No later report is available so far as the present author is aware. Cultivated white rice when buried in the soil at the depth of ordinary ploughing loses its viability during the first winter in California (Goss & Brown, 1939). Italian and California red rices behave in a similar way, though they are slightly more resistant to burial. In general, the seed remained alive longer in the irrigated than in the non-irrigated plots. Under dry storage at the soil temperature conditions existing in California, all of the red rices tested showed good viability after three winters. This experiment was terminated after ten years, when some germination was still obtained from four of the rice lots stored in irrigated soil in Arkansas (Goss & Brown, 1940).

Arai & Kataoka (1956) found a seasonal variation in the viable seed

9

population of *Alopecurus aequalis*, a dominant weed in fields of wheat and barley, and in its vertical distribution in the soil. The density of the viable seed populations in top-soils decreased daily after the field had been drained in autumn, the decrease varying widely with the depth of the soil layer and with tillage.

E. O. Brown & Porter (1942) buried weed seeds in the soil at different depths and tested their germination after three years. The viability of seeds of *Lepidium draba* L., *Lepidium repens* L., and *Hymenophysa pubescens* C.A.Mey, deteriorated rapidly when buried in soil at a depth of 4 to 6 in. Seeds of *Euphorbia escula* L. kept well at 6- and 18-in. depths. *Centaurea calcitrapa* L. and *C. solstitialis* L. had a short life-span in the soil, declining from 90 to 10 per cent germination within three years. Seeds of *C. repens* Boiss. were less subject to deterioration. Dormancy in *Solanum* seeds present at the beginning of the test disappeared after three years of burial, but the actual germination of seeds of *S. elaeagnifolium* Cav. was greatly impaired by three years' burial at 18 in. Ninety-eight per cent germination of *S. carolinense* L. seeds was obtained after burial for three years at a depth of 4 to 6 in., but very few survived burial at 16 to 18 in. for the same period.

One of the remarkable things about the longevity of weed seeds in the soil is that most of them do not have impermeable coats and hence absorb water immediately upon exposure to moist soil. That seeds which have imbibed their full complement of water should remain viable over such a long period is noteworthy. Furthermore, most of these seeds can scarcely be considered dormant, though some of them may develop secondary dormancy. That they do not develop a 'deep' dormancy, when buried in the soil, is demonstrated by the fact that when the soil in an ordinary garden plot, for example, is disturbed by cultivation, many weed seedlings appear even if care has been taken to remove all such plants for several preceding years. Exposure to light, alternating temperature, mechanical disturbance, or some other unknown factor may serve as a stimulus to bring about germination of these buried viable seeds. Furthermore, it has been shown by Kjaer (1948) that some seeds actually remain viable longer in the soil than in dry storage. His report includes tests of up to a ten-year period at the end of which nine of the different seeds that germinated after burial failed to germinate after dry storage. Out of the seventeen stored, only one species, *Geranium dissectum*, survived the dry storage but failed to germinate after burial in soil. The other seven species failed to germinate after ten years regardless of the storage condition. Other evidence of the superiority of moist soil over dry storage at room temperature is to be found in various reports (Crocker, 1948, p. 40).

Seeds of the Western bitterweed, *Actinea odorata*, were stored in the open, fully exposed to weather conditions, and in the laboratory in Texas (Reynolds *et al.*, 1938). Both lots of seeds had a viability of 72 per cent at the beginning of the test. Both lots lost their germinating power with increasing age, but the seeds outside deteriorated more rapidly than those in the laboratory. All of the latter were, however, dead after five years of storage.

MOIST STORAGE OF SEEDS OF *Amaranthus retroflexus* L.

When seeds take up water preparatory to germination, their metabolic activities become greatly accelerated. It is evident that seeds which have imbibed water could not remain viable for very long in the soil unless definite curtailment of these activities took place. In an effort to obtain more definite information on the effects of burial, freshly harvested seeds of the pigweed, *Amaranthus retroflexus* L., were used (Barton, 1945*b*). Seeds of this plant when freshly harvested show a degree of dormancy which varies from year to year and, with seeds collected in various localities, in any one year. The 'dormancy' is expressed in the requirement of a high temperature for germination. The best temperature for germination of freshly harvested seeds was 35 °C., at which 86 per cent of seedlings were obtained from one lot and 73 per cent from another lot immediately after collection. Furthermore, the temperature was quite specific at that time. Some germination occurred at 25° and 30°C., but the percentages here did not exceed 18 in either lot. Only occasional seedlings appeared over the rather wide range of other temperatures tried.

When these seeds were stored dry after harvesting, a gradual change in their capacity for germination took place, so that after two or three months they germinated over a wide range of temperatures including 20°C. If, however, they were placed on a moist medium at 20°C. immediately following harvesting, this after-ripening did not proceed but, rather, primary dormancy was maintained or a secondary dormancy developed. This resulted in a different pattern of germination behaviour.

Careful records of the germination on moist glass-wool at 20°C. of four lots of *Amaranthus retroflexus* seeds have been kept for eight years (Barton, 1945*b*). All four lots were similar in germination behaviour. A summary of the germination of two different collections made in 1942 is shown in Table III. These seeds have exhibited a periodicity in germination which has been remarkably uniform for the four lots. It may be noted that the data in collection A are based on twenty-two lots of 2,500 seeds each (55,000 seeds), whereas those in collection B are based on eighteen lots of 2,500 seeds each (45,000 seeds). Germination behaviour of individual lots of 2,500 seeds was recorded, but all lots of a single collection have been combined in the figures shown in Table III. The number of seedlings appearing by the end of each two-month period during each of eight years is given. Collections were made in October, so that the first two months represent October and November and the last two months represent August and September. Within a period of two months after the seeds had been moistened and placed at 20°C., a total of 186 seedlings had been produced from seeds of collection A and 378 from those of collection B. A strong germination-pulse was noted for the tenth to twelfth months of storage during the first and succeeding years up to eight. This periodicity of germination is apparently independent of external conditions, for it took place under nearly constant conditions. This indicates a varying degree of the primary dormancy or the induced secondary dormancy of the individual seeds of the original lot. Moist *Amaranthus* seeds, held without germination at 20°C., could be induced to germinate at that same temperature by rubbing, by drying for from three hours to three days, or by exposure to 35°C. for twelve to twenty-four hours. Germination also proceeded immediately after removal to higher constant or alternating temperatures.

TABLE III

Amaranthus retroflexus, 1942 CROP. GERMINATION AFTER VARIOUS PERIODS DURING EACH OF 8 YEARS ON MOIST GLASS-WOOL AT 20°C.

Collection	Yrs. of moist storage	No. of seedlings produced each 2-monthly period during the year						
		2 (Oct.-Nov.)	4 (Dec.-Jan.)	6 (Feb.-Mar.)	8 (Apr.-May)	10 (Jun.-Jul.)	12 (Aug.-Sept.)	Total
	1	186	31	15	47	12,201	1,223	13,703
	2	290	24	22	831	12,814	4,227	18,208
	3	2,993	378	707	237	401	738	5,454
A	4	108	12	147	270	513	113	1,163
55,000	5	87	6	3	22	106	66	290
seeds	6	6	0	5	151	454	47	663
	7	2	1	0	9	36	171	219
	8	1	3	0	1	50	74	129
	Total	3,673	455	899	1,568	26,575	6,659	39,829
	1	378	82	72	57	12,763	3,155	16,507
	2	82	25	5	24	184	173	493
	3	2,994	135	46	1,132	432	1,308	6,047
B	4	91	23	79	744	5,638	435	7,010
45,000	5	216	60	33	213	4,829	973	6,324
seeds	6	34	0	1	139	1,732	452	2,358
	7	2	0	0	8	75	253	338
	8	5	12	0	11	2,381	915	3,324
	Total	3,802	337	236	2,328	28,034	7,664	42,401

Gaseous exchange of *Amaranthus retroflexus* seeds, measured at intervals of from o to 901 days of moist storage at 20°C., showed at least a ten-fold reduction in respiration. The beginning of this reduction became apparent very early (after two days), and was definite after eight days in moist storage. The decreased respiratory rate is probably an indication of the reduced metabolic activity of seeds in the soil which have imbibed water, though it is believed that further reduction from that measured here would have to take place to permit these seeds to live for as long as forty years. The present experiments were discontinued after eight years.

Toole & Toole (1954) found that seeds of seven lettuce varieties which had imbibed their full complement of water remained viable on moist filter-paper for 105 days, whereas dry seeds in a humid atmosphere were dead after forty-two or twenty-one days, depending on the variety. Storage was at 30°C., a temperature known to maintain dormancy of moist lettuce seeds.

Additional studies of respiration as well as of other processes taking place within the seed are needed for an explanation of the physiological responses of buried seeds.

In the case of seeds with impermeable coats, it is obvious that their germination will proceed in soil 'storage' as soon as micro-organisms or mechanical abrasion makes the coats permeable to water and sufficient moisture is available. The several layers of the seed-coat of *Nelumbo* are shown in Plate 1(*b*), with the parts which prevent water absorption indicated. In the old seeds that had been in the soil for hundreds of years, the epidermis and much of the outer ends of the palisade cells had disappeared, but most of those old seeds which were recovered from the soil still required coat treatment for germination (Ohga, 1926). This probably indicates simply that such old seeds as became permeable over the years germinated, but, because they were buried at such great depths, the resultant seedlings failed to survive.

III

MOISTURE EFFECTS

MANY factors, such as moisture, temperature, gaseous exchange, seed-coat character, maturity, microflora and insect infestation, may determine the longevity of seeds under natural or controlled storage.

It has been well established by many investigators that the moisture content of seeds is of the utmost importance as a factor in the determination of their longevity. This applies not only to the absolute moisture content but also to fluctuation in moisture content—especially to fluctuations around the 'critical' moisture content which varies according to the type of seed. That temperature and gaseous exchange are of vital importance in connection with the keeping quality of seeds has also been demonstrated repeatedly and will be discussed below. None of these items can be considered alone, as the effect of one depends upon others.

The importance of moisture in the storage of seeds has been recognized by the International Seed Testing Association in the formation of a Committee on Seed Moisture Content and Seed Storage. This Committee obtains information on seed moisture and seed storage studies made in different areas of the world and serves as a clearing house for such information (E. H. Toole, 1957). As early as 1934, the appearance of abnormal seedlings which caused difficulty in the interpretation of germination tests was attributed to high moisture content of the seeds at some time after harvest (E. Brown *et al.*, 1934).

In view of these facts, it becomes highly desirable to ascertain the actual amounts of water which will be absorbed by seeds in different localities, i.e. under different conditions of humidity and temperature. Such knowledge would make it possible to determine whether air-dry seeds, in any particular region, could be stored with safety. Data on moisture contents of six varieties of seeds, stored at four different temperatures and three different relative humidities, have been presented by Barton (1941). Correlations with the germination capacity under all of these conditions have been made. As the results of these tests have a direct bearing on keeping quality of seeds in general, they will be presented here in some detail.

Seeds used were of lettuce (*Lactuca sativa* L.), onion (*Allium cepa* L.), tomato (*Lycopersicum esculentum* Mill.), flax (*Linum usitatissimum* L.), peanut (*Arachis hypogaea* L.), and long-leafed pine (*Pinus palustris* Mill.).

TABLE IV

MOISTURE CONTENTS OF SEEDS STORED AT VARIOUS TEMPERATURES AND RELATIVE HUMIDITIES
(MOISTURE EXPRESSED AS PERCENTAGE OF DRY WEIGHT OF SEEDS)

Per cent moisture after storage for 8, 29, 105, and 372 days

Seed+per cent moisture at time of storage	Per cent Rel. hum.	5°C				10°C				20°C				30°C			
		8	29	105	372	8	29	105	372	8	29	105	372	8	29	105	372
Lettuce 6·5	35	7·0	6·8	5·8	6·2	8·4	6·8	5·9	6·2	6·6	6·1	5·2	5·8	—	5·3	4·5	5·0
	55	7·5	9·0	8·0	8·7	10·2	10·0	8·5	9·1	9·8	8·5	6·8	7·4	8·6	6·5	5·4	5·9
	76	8·3	9·6	10·5	12·7	10·1	13·3	19·6	15·7	10·2	12·3	11·1	11·8	8·3	10·0	10·4	10·7
Onion 10·2	35	10·7	10·8	9·2	9·9	12·1	10·6	9·1	10·1	10·5	9·7	8·3	9·1	9·1	8·5	7·2	8·0
	55	11·8	13·7	12·4	13·4	15·1	15·4	15·1	14·0	15·6	13·6	11·0	11·8	12·7	10·4	8·8	9·6
	76	12·8	15·1	15·2	17·3	13·7	18·5	30·5	18·7	14·4	16·3	17·1	17·5	12·2	14·4	13·8	15·0
Tomato 8·9	35	9·3	9·2	8·1	8·9	10·6	9·8	8·3	8·5	9·3	8·5	7·4	8·3	8·1	7·5	6·3	7·1
	55	10·2	12·2	11·4	11·7	15·7	14·4	13·1	13·6	12·8	11·7	9·7	10·3	11·7	9·3	7·7	8·5
	76	11·2	12·9	13·8	15·5	17·1	17·3	15·7	18·5	12·8	14·1	15·9	15·4	10·8	12·6	12·6	13·1
Flax 7·6	35	8·0	7·8	6·8	7·4	9·5	7·6	6·8	7·4	7·6	7·0	6·1	6·7	6·7	6·3	5·3	5·9
	55	8·6	10·4	9·3	9·9	12·6	12·0	11·0	10·8	11·0	9·7	8·1	8·7	9·8	7·7	6·4	7·0
	76	9·5	11·5	11·4	14·2	10·8	14·5	16·5	20·4	10·6	12·1	13·0	12·5	9·6	11·1	11·2	11·6
Pine 8·1	35	8·5	8·6	7·0	7·7	9·2	8·5	7·4	7·8	8·3	7·4	6·3	6·9	7·1	6·4	5·5	6·2
	55	8·9	10·5	9·8	10·9	11·1	12·9	10·8	11·2	10·9	10·1	8·4	9·2	10·3	8·2	6·9	7·5
	76	9·2	11·1	11·6	13·9	9·9	11·6	14·6	16·0	10·9	12·7	12·8	13·4	9·8	11·6	11·8	12·3
Peanut 5·8	35	6·0	6·1	4·8	—	7·1	6·0	4·8	5·9	5·9	5·1	4·5	4·8	4·9	4·6	3·9	—
	55	6·3	7·7	7·0	7·4	8·0	8·5	7·7	8·1	8·4	7·7	5·9	6·1	7·6	5·6	4·6	—
	76	6·7	8·6	8·6	11·1	7·2	9·6	13·9	12·4	8·5	10·0	9·1	10·6	7·2	8·7	9·7	—

Large desiccators containing saturated solutions of magnesium chloride ($MgCl_2$. $6H_2O$), calcium nitrate ($Ca(NO_3)_2$. $4H_2O$), or sodium chloride (NaCl), with an excess of the salts, were placed in constant temperature chambers of 5°, 10°, 20°, and 30°C. These three salt solutions provided atmospheres of approximately 35, 55, and 76 per cent relative humidities, respectively (H.M. Spencer, 1926). The seeds were placed in muslin bags over the solutions in the desiccators. Moisture determinations were made on each seed-lot after 8, 17, 29, 43, 105, 150, 232, and 372 days, by drying in a vacuum oven at 78°C. for forty-eight hours. All tests for the germination capacities of the seeds were made in ovens at controlled temperatures which had previously been determined as favourable.

Results of moisture determinations made 8, 29, 105, and 372 days after storage under special conditions are shown in Table IV. Tests made after eight and seventeen days on lots representing all of the different conditions showed fluctuations which indicated that the seeds had not reached an equilibrium with the moisture in the storage chamber. Tests made as long as twenty-nine days after storage showed that an equilibrium was being established by that time, and changes in moisture content of individual samples observed on subsequent dates were doubtless due in great part to the error of sampling.

It was noted that the seeds showed differential water absorption according to the species, and that the amount of moisture taken up under all conditions by each seed type always appeared in the same relative position. This is shown in Fig. 2, where the moisture determinations made forty-three days after storage are used. In the order of increasing water-absorption capacity, the seeds were of peanut, lettuce, flax, pine, tomato, and onion. This order persisted regardless of storage temperature and atmospheric humidity.

The chemical composition of the seeds may account, at least in part, for this variation in the amount of moisture absorbed by different seeds under identical conditions. Peanut, flax, lettuce, and pine seeds, which contain large amounts of oil, have low retention capacities. On the other hand, tomato and onion seeds, both of which contain larger amounts of protein and smaller amounts of oil, occupy intermediate and high positions as regards ability to absorb moisture. Ewart (1896) stated that the resistant power of a seed to desiccation depended partly on the nature and thickness of the coat and partly on the form in which the reserve food was stored. Other conditions being similar, albuminous seeds were least resistant to desiccation, oily seeds next, and starchy seeds most resistant. The present tests confirm these statements for albuminous and oily seeds. The types of seed-coats, also, doubtless play a role.

It should be pointed out that, at the beginning of the tests, seeds of the same species possessed the same viability in all storage conditions; but, as the storage period lengthened, loss of viability progressed much more rapidly under conditions of high humidity and temperature. This

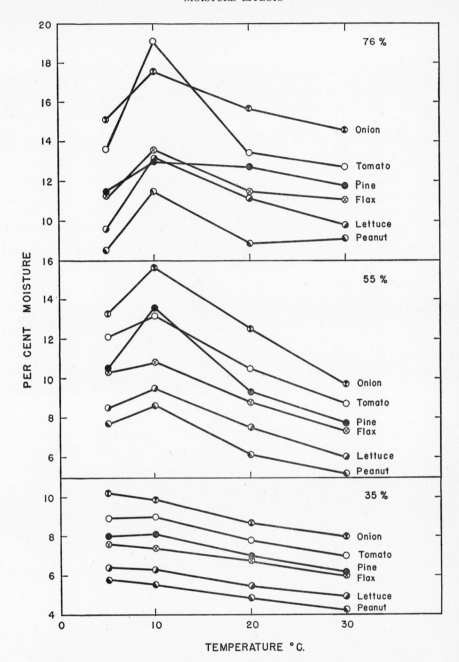

FIG. 2.—Graphs indicating moisture content of seeds after forty-three days of storage at various temperatures and at relative humidities of 76, 55, and 35 per cent.

means, then, that moisture determinations were made with increasing proportions of non-viable seeds. That the state of viability was without effect on the actual amount of water absorbed, was demonstrated by the failure of the moisture content to change in relation to decreased germination capacity. This may be seen from a comparison of Tables IV and V. Other workers have also shown that seed viability does not affect the total water absorption. Heinrich (1913) weighed seeds of *Lolium perenne* after 0 to 384 hours in a moist room at 20°C. and found that living seeds contained more water than dead ones up to twenty-four hours, after which they were approximately the same in this respect. Atkins (1909) found that seeds, whether living or dead, take up the same quantity of water in their initial stages and that there is no difference until germination begins, and Simpson & Miller (1944) found essentially the same moisture contents in living and dead seeds.

With exposure to a relative humidity of only 35 per cent, seeds took up approximately the same amount of water at 5° and 10°C. (Barton, 1941), but in every case they remained drier at the higher temperatures of 20° and 30°C. (Fig. 2). At 55 and 76 per cent relative humidities, however, the peak of moisture absorption was at 10°C. and the least absorption was generally at 30°C. There was no indication that equilibrium was reached more quickly at 30°C. than at any of the other temperatures tried. Boswell *et al.* (1940) stored seeds of bean, sweet corn, beet, cabbage, carrot, onion, spinach, tomato, and Spanish peanut at 50° and 80°F. at relative humidities of 78 to 81 per cent, 66 per cent, and 44 to 51 per cent, and reported that at a given relative humidity the seeds developed slightly higher moisture contents at 50° than at 80°F. These authors stated that this could not be explained by the physical characteristics of the air, such as differences in vapour pressure, weight of water in the air, or vapour deficit.

It has been reported that seeds of blue lupine and Austrian winter field pea absorb moisture more rapidly at 35° than at 20°C. (McKee & Musil, 1948). According to A. J. M. Smith & Gane (1939), the rate at which moisture is taken up or lost by a certain kind of seed depends upon the relative humidity of the air and the thickness of the mass of seeds. They studied the hygroscopic properties of Chewing's fescue grass seeds after drying. The water content of wheat grain at a given relative humidity is less at 25°C. than at 10°C. (Gane, 1941). Gane (1948) also found a difference in water content between different seeds exposed to the same relative humidity, and a decrease in water content with increased temperature at a constant relative humidity.

D. A. Coleman & Fellows (1925) determined moisture contents of cereal grains and flax seeds exposed to 15, 30, 45, 60, 75, 90, and 100 per cent relative humidity at 25°C., but no other temperature was used.

Barton (1941) showed that there were significantly higher moisture contents of seeds at a given relative humidity at 10°C. than at 30°C., in spite of the greater amount of moisture available in the atmosphere at the higher temperature. Less moisture was absorbed at 20°C. than at

TABLE V

GERMINATION PERCENTAGES OBTAINED FROM SEEDS STORED AT VARIOUS TEMPERATURES AND RELATIVE HUMIDITIES

Per cent germination after storage for 8, 43, 150, 232, and 372 days

Seed + per cent germination at time of storage	Per cent Rel. hum.	5°C.					10°C.					20°C.					30°C.				
		8	43	150	232	372	8	43	150	232	372	8	43	150	232	372	8	43	150	232	372
Lettuce 63	35	59	79	71	73	67	51	78	66	70	53	56	69	63	59	32	46	68	33	41	2
	55	50	80	78	75	50	44	69	65	57	22	45	62	51	43	3	41	51	16	6	0
	76	50	67	67	68	36	57	64	1	7	0	58	46	1	0	0	52	4	0	0	0
Onion 66	35	56	67	62	57	55	38	63	47	53	35	58	61	45	32	29	54	47	16	13	15
	55	51	70	69	58	53	62	50	33	29	16	41	15	5	9	3	17	11	2	1	4
	76	41	67	59	44	27	55	62	30	17	13	51	15	1	1	1	33	3	0	1	0
Tomato 93	35	94	88	91	91	94	88	91	92	92	91	87	90	93	88	90	90	89	92	88	91
	55	94	93	92	91	90	90	91	93	86	89	86	85	89	96	89	94	89	86	87	83
	76	89	92	89	89	88	92	87	88	75	76	92	83	83	73	45	89	86	32	0	0
Flax 97	35	97	97	89	98	98	91	94	95	95	92	93	91	95	95	86	93	97	86	91	83
	55	96	97	88	95	96	96	97	96	93	85	94	94	90	87	90	96	98	90	79	74
	76	92	96	92	88	94	94	97	78	79	71	98	96	78	60	32	94	94	1	0	0
Pine 96	35	69	72	54	58	—	83	59	56	35	—	76	82	73	67	—	82	45	31	19	—
	55	76	71	71	70	—	87	50	47	24	—	84	40	28	24	—	60	11	0	0	—
	76	90	59	52	41	—	87	49	15	3	—	61	26	0	0	—	68	1	0	0	—
Peanut 96	35	98	100	98	100	—	98	100	96	100	—	100	100	98	100	100	100	100	98	97	—
	55	98	100	94	100	—	98	100	98	96	—	96	98	98	100	100	98	100	98	98	—
	76	100	100	100	100	—	94	98	96	96	—	96	98	82	42	—	98	96	0	0	—

— Indicates that no test was made.

19

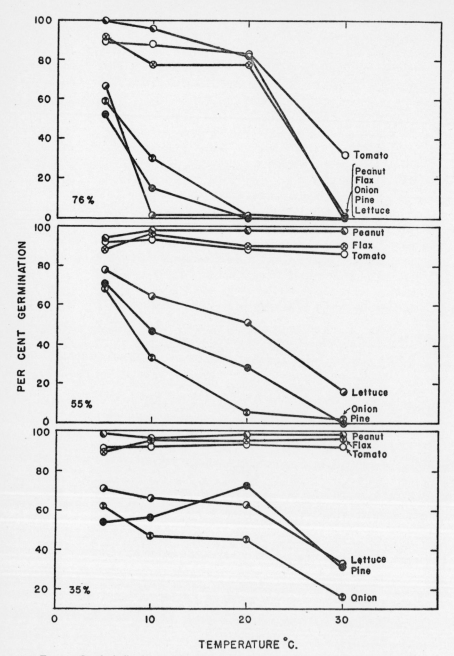

FIG. 3.—Graphs indicating germination after storage for 150 days at various temperatures and relative humidities.

10°C., but more than at 30°C. Approximately the same amount was taken up at 5° and 10°C. at 35 per cent relative humidity but, as the relative humidity was increased to 55 and to 76 per cent, seeds absorbed more water at 10°C. than at 5°C. The explanation of these phenomena must be sought, not only in the physical conditions of the atmosphere surrounding the seeds but also in the physical conditions of the seeds themselves. The actual determination of the vapour pressures of both seeds and atmosphere at the different temperatures used, and the relations of these figures to each other, may provide a basis for explanation of the behaviour.

Viability tests made by Barton (1941) after 150 days of storage are shown in Fig. 3. A comparison of these graphs with those in Fig. 2 reveals that the actual amount of water absorbed by the seed is not directly related to its loss of viability under any given set of conditions. The three seed types exhibiting greatest retention of viability under all conditions were peanut, flax, and tomato. On the basis of the actual amount of moisture absorbed these seeds occupy low, intermediate, and high positions, respectively (Fig. 2). Similarly, lettuce, pine, and onion seeds, which lose viability most readily under all conditions, have low, intermediate, and high moisture contents, respectively. Also, deterioration does not appear to depend upon composition of the seed.

TABLE VI

ESTIMATED MAXIMUM SAFE SEED MOISTURE CONTENTS FOR STORAGE
FOR ONE YEAR AT DIFFERENT MEAN TEMPERATURES OF
STORAGE (approximate guide only)

Kind of seed	Maximum safe seed moisture content for average temperature of storage indicated			Kind of seed	Maximum safe seed moisture content for average temperature of storage indicated		
	4·5-10°C. 40-50°F.*	21°C. 70°F.	26·5°C. 80°F.		4·5-10°C. 40-50°F.*	21°C. 70°F.	26·5°C. 80°F.
	per cent	per cent	per cent		per cent	per cent	per cent
Bean, Kidney	15	11	8	Okra	14	12	10
Bean, Lima	15	11	8	Onion	11	8	6
Beet	14	11	9	Pea, Garden	15	13	9
Cabbage	9	7	5	Peanut			
Carrot	13	9	7	(shelled)	6	5	3
Celery	13	9	7	Pepper	10	9	7
Corn, Sweet	14	10	8	Spinach	13	11	9
Cucumber	11	9	8	Tomato	13	11	9
Lettuce	10	7	5	Turnip	10	8	6
				Watermelon	10	8	7

* Special precautions needed when removed to higher temperature.

Boswell *et al.* (1940) also reported that among different kinds of seed, deterioration was not always correlated with the relative moisture-absorbing capacity of the seeds.

E. H. Toole (1939) states that the relative loss of viability of different kinds of seeds at a given temperature and air humidity depends in part on the moisture content and in part on the nature of the seeds. Tomato seeds will retain viability longer than onion seeds at a given temperature and air humidity, but the longevity is fairly comparable at a given temperature when the moisture content of the seed is the same. On the other hand, peanuts take up comparatively little moisture, but they deteriorate rapidly at high temperatures even with a low moisture content. Maximum safe seed-moisture contents for storage for one year at different storage temperatures have been estimated by E. H. Toole (1942) and are given in Table VI.

MOISTURE RELATIONSHIPS OF SEEDS STORED 'OPEN' IN THE LABORATORY

Whether 'air-dry' seeds contain safe moisture contents for storage depends entirely upon the locality and time of year that storage is to be accomplished.

In order to determine more exactly what fluctuations actually take place under ordinary atmospheric conditions, during the course of a year at Yonkers, New York, moisture determinations were made in February, May, July, August, September, November, and December 1939, and January 1940, of seeds of carrot (*Daucus carota* L. var. *sativa* DC.), eggplant 'aubergine' (*Solanum melongena* L.), lettuce, tomato, and long-leafed pine stored open in the laboratory. The results of these tests showed that the moisture content of seeds in August was practically double that in January or February. This applied to every type of seed tested, although there was considerable variation in the actual amounts of water contained in the different varieties. These differences are shown graphically in Fig. 4, where a comparison of lettuce, pine, and tomato seeds is made. In spite of the differential water-absorbing capacities of these seeds, all of them contain the maximum moisture in August and are relatively dry during the months of November, December, January, and February. The moisture contents of carrot and eggplant seeds at the various testing periods were practically the same as for tomato seeds. Pine seeds took up less moisture from the air, while lettuce seeds remained driest.

This test simply serves to emphasize the necessity of drying seeds to a known moisture content before storage, as the amount of moisture present in 'air-dry' seeds depends not only on the locality but also on the time of year. 'Air-dry' seeds placed in sealed containers during August, in Yonkers, New York would contain twice as much moisture as the same seed-lot sealed during the winter months. In the evaluation of sealed as opposed to open storage, these facts should be considered.

The differential absorption of water by the different species furnishes additional evidence that the 'critical' moisture content varies with different seeds. It is believed that fluctuation in moisture content is most deleterious to seed viability when it occurs around the 'critical' point.

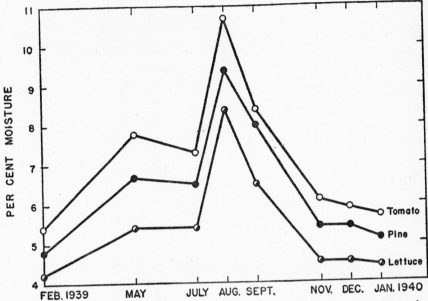

FIG. 4.—Graph showing moisture contents at various times of the year of seeds stored 'open' in the laboratory. Moisture is expressed as percentage of dry-weight of seeds. Duplicates of approximately 2 gm. each were used for the tests.

The importance of fluctuation in moisture content on the keeping quality of dandelion (*Taraxacum officinale* Weber) seeds has been investigated by Barton (1939*b*, 1953), who stored the seeds at three temperatures: laboratory, 5°C., and −4°C. Air-dry seeds at the time of storage contained 7·9 per cent moisture. Some samples were stored with this moisture content, while other samples of the same seed-lot were adjusted, by storing in desiccators over calcium oxide, to 6·2, 5·0, and 3·9 per cent moisture before storage in sealed containers. Sealing was accomplished in two ways. In one case, lots of approximately 400 seeds each were placed in small glass tubes hermetically sealed in air and in a vacuum; in the other case, larger lots were placed in tin cans with tight-fitting lids sealed with sealing wax. Each tin can was opened and samples were taken out at intervals for testing, after which the cans were re-sealed. Thus the seeds remaining after each germination test had been subjected to fluctuation in moisture and gaseous exchange attendant upon the opening of the can—a condition which did not exist for individual testing samples stored in glass tubes. The method of sealing proved important in keeping-quality as shown by Fig. 5, which depicts the results of extended storage for up to sixteen years of seeds with initial moisture contents of 7·9 and 3·9 per cent. Seeds with 7·9 per cent moisture (Fig. 5(A)) stored in a tin can in the laboratory lost germination capacity completely within two years, while those stored in sealed glass tubes still gave 66 per cent germination at the end of that time. Furthermore, 12 per cent of the latter germinated after four years' storage. An even more striking difference in the effects of two sealing methods is to be seen in the 5°C. storage. In tin cans a high germination capacity was retained for four years, after which there was rather rapid deterioration up to ten years, when only 7 per cent of the seeds were alive. In glass tubes at 5°C., on the other hand, practically full viability was retained for fourteen years, at which time the supply of seeds was exhausted. Even at the more favourable storage temperature of −4°C., the advantage of the glass tube over the tin can for keeping these seeds is evident, though the 'break' in germination representing the beginning of deterioration in this case was delayed until after twelve years of storage.

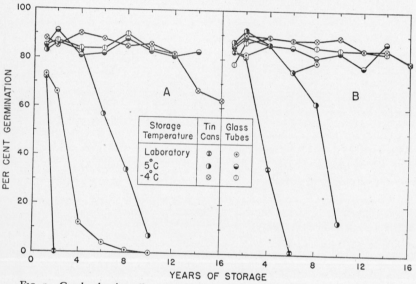

Fig. 5.—Graphs showing effect of temperature and method of sealing on the viability of stored dandelion seeds with an initial content of (A) 7·9, and (B) 3·9 per cent moisture. Germination at time of storage was 91 per cent.

The life-span of dandelion seeds may be extended by drying to a 3·9 per cent moisture content before storage as shown in Fig. 5(B). In this case, deterioration in a tin can in the laboratory became marked only after four years of storage, but the favourable effect of storing in glass tubes is still evident. Similar effects are to be seen for 5°C. storage, but they were not evident for −4°C. storage up to sixteen years.

The more rapid deterioration of seeds stored in sealed containers which are repeatedly opened and re-sealed, as compared with seeds in sealed containers which are never opened from the time of storage to the time of testing, sheds some light on the moisture relations of seeds.

Evidence in the form of delayed ill-effects of opening and sealing, by decreased initial moisture content, of the seeds of dandelion leads to the conclusion that fluctuation in moisture content is harmful. Reference is made again to Fig. 5 on this point. Moisture determinations, made by drying the seeds in a vacuum oven at 78°C., have shown that seeds with a moisture content of 7·9 per cent at the beginning will gradually dry out to about 6 per cent with repeated opening of the storage can. As far as moisture content itself is concerned, this should be an advantage, but the fact is that dandelion seeds 'keep' better with a constant moisture content of 7·9 per cent than with a fluctuation downwards. Deterioration of seeds with 3·9 per cent moisture is accelerated when repeated opening of the can has brought the moisture content up to about 6 per cent as shown by actual moisture determinations. Dandelion seeds with intermediate initial moisture contents (6·2 and 5·0 per cent) were intermediate in their responses to moisture fluctuation.

It should not be assumed from this discussion that it is always better to seal seeds with high moisture content than to leave them open. If the moisture content is excessive, there is increased injury in sealed containers; but within certain limits, which probably vary for each kind of seed, fluctuation in moisture content results in greater damage to the seed than could normally be expected from constant levels of moisture content.

Subsequent to the findings just described, a test was set up designed to include controlled fluctuation of humidity at a constant temperature of 20°C. (Barton, 1943). Desiccators with different relative humidities were used as storage chambers, a constant humidity being maintained in each desiccator by means of a saturated salt solution with an excess of the salt (H. M. Spencer, 1926). Fluctuations in relative humidity between 35 and 55 per cent, between 35 and 76 per cent, and between 55 and 76 per cent, at intervals of two, four, and eight weeks, were made by transferring the bags containing the seeds from one desiccator to another. As controls, certain lots were left at each of the constant humidities for the entire storage period. The seeds used were of onion (*Allium cepa* L.), dandelion (*Taraxacum officinale* Weber), eggplant (*Solanum melongena* L.), and tomato (*Lycopersicum esculentum* Mill.), the first two being tried because they are very sensitive to adverse storage conditions and the second two because they are relatively resistant to the same conditions. Viability tests were made on moist filter paper in ovens at controlled temperatures, except in the case of dandelion which was germinated in the laboratory.

In general, when alternation from one humidity to another was effected, the variation in moisture content of the seeds tended to be the same as was indicated by the determinations made from constant-humidity storage. In some cases, especially if the alternations were made every two weeks, variation in moisture content of the seeds was somewhat less than that indicated by the determinations made from constant-humidity storage. Seeds transferred from a high to a low humidity failed to reach the low moisture content expected before they were transferred again to a higher humidity, where again they failed to develop the maximum moisture content to be expected.

Onion seeds held at constant relative humidities of 35, 55, or 76 per cent acquired moisture contents which averaged 8·6, 11·4, and 14·9 per cent, respectively, and remained at these values throughout the test period. Onion seeds which were alternated between 35 and 55 per cent relative humidity attained an average of 9·4 per cent moisture after two weeks at 35 per cent and 11 per cent after two weeks at 55 per cent, so that the average variation in moisture content was 1·6 per cent every two weeks. Similarly, the average variation when the seeds were changed from 35 to 76 or from 55 to 76 per cent moisture every two weeks was 4·7 and 2 per cent, respectively.

Except at a constant relative humidity of 76 per cent, neither eggplant 'aubergine' nor tomato seeds showed any reduction in viability under any of the storage conditions up to the end of sixty-four weeks, at which time the experiment was terminated. These seeds are known to be resistant to unfavourable storage conditions. A constant relative humidity of 35 per cent imparted a moisture content of 9 per cent to onion seeds and permitted the retention of their viability for one year at 20°C., with the percentage germination only slightly reduced at the end of that period. At 55 per cent relative humidity, at which onion seeds came to contain about 11 per cent of moisture, serious deterioration occurred after twenty weeks of storage. At 76 per cent relative humidity considerable loss in germinative power was evident after eight weeks of storage, and, after twelve weeks at this humidity, where the moisture content of the seeds was about 15 per cent, onion seeds yielded only 23 per cent germination as compared with 96 per cent at the beginning of the special storage period.

These comparative effects of low, medium, and high humidity storage chambers were to be expected, as previous work had already demonstrated similar characteristics in onion seed. But the question remains as to whether alternation of humidities in the storage chambers, and hence fluctuations of moisture content within the seed itself, is more deleterious than a constant high moisture content. At first it appeared that the answer to this question was in the negative. Seeds which had two-weekly, four-weekly, or eight-weekly alternations from 35 to 55 per cent relative humidity remained viable longer than those at a constant relative humidity of 55 per cent, and deteriorated more rapidly than those at a constant relative humidity of 35 per cent. Seeds kept under conditions of alternation of humidity, then, were intermediate in germination capacity between those kept at the two humidities concerned. This same response applied to seeds alternated between 35 and 76 per cent relative humidity and between 55 and 76 per cent relative humidity. However, when these last two alternations were made eight-weekly, the seeds deteriorated as rapidly as they did when kept constantly at the higher humidity. This was noteworthy in view of the fact that the seeds had been kept for half of the time at either 35 or 55 per cent relative humidity, both of which permit retention of viability for fairly long periods, and consequently had remained for only half as long at the very harmful 76 per cent relative humidity as the control lot at this humidity.

Further examination of the deterioration effects after certain storage periods revealed that still other factors were involved. The data showed that during the first twelve weeks of storage at two- or four-weekly alternations of either 35 to 76 or 55 to 76 per cent relative humidity, the lower humidities tended to prevent rapid loss of viability during the periods at 76 per cent. However, at the sixteen-week storage period and thereafter, the lower humidities could no longer counteract the deleterious effect of the high humidity, and the germination-capacity curves more nearly approached that of the constant high humidity. In other words, at sixteen weeks of storage and thereafter, deterioration at alternating humidities was at a more rapid rate than would be expected from the behaviour at each of the humidities. This was especially marked for the alternation of 35 to 76 per cent.

On the other hand, at an alternation of 35 to 55 per cent, the periods at the favourable relative humidity of 35 per cent were sufficient to decrease the deterioration rate from that expected on the basis of the behaviour at 55 per cent for up to forty-eight weeks of storage—after which the same effect as mentioned above was indicated by tests made after fifty-two, fifty-six, and sixty-four weeks of storage. This delay in the manifestation of the harmful effect produced by alternation of moisture content was no doubt due to the fact that 55 per cent relative humidity was much less harmful to germination capacity than was 76 per cent.

The eight-weekly alternations deserve special consideration. The first viability test of these seeds was made after sixteen weeks of storage. As all seeds in alternating humidities were placed in the lower humidity first, this means that seeds at 55 to 76 per cent alternation, for example, had the first eight weeks at 55 per cent (which permits high retention of viability) and the second eight weeks at 76 per cent. In spite of the fact that these seeds had only eight weeks of the unfavourable 76 per cent relative humidity, after which, if no other factors were involved, they should still have given 67 per cent germination, they gave only 14 per cent germination. A similar effect was obtained after eight-weekly alternation of 35 to 76 per cent relative humidity, when the germination obtained was 29 per cent.

When the total length of time at the higher humidity—whether that time was obtained by alternation or by constant humidities—is considered, the germination percentages parallel the length of time except in the case of the eight-weekly alternation which gives a much lower germination percentage as described above. The total period at high humidity, then, is of great importance in determining the length of life of onion seeds, but intervening periods of lower and hence more favourable humidities do not prolong the life-span as much as would be expected from the behavioural data of seeds kept constantly at these lower humidities.

If the total period at 76 per cent relative humidity in alternating humidities were the only factor involved, two-weekly, four-weekly, and eight-weekly alternations should bring about the same deterioration when the total time at 76 per cent is the same. Also, an alternation of 35 to 76 per cent relative humidity should be as deleterious as one of 55 to 76 per cent. Such was not found to be the case. Therefore, some other factor must have been operating, and it appears from the data at hand that this factor was fluctuation in moisture content.

Dandelion seeds were similar to onion seeds in their response to moisture fluctuation (Barton, 1943). The length of time at a harmful relative humidity of 76 per cent was directly related to the deterioration rate of dandelion seeds, regardless of the storage conditions before and after this exposure.

Artificial Drying

As the moisture content of seeds is of prime importance in determining their keeping quality, and as most seeds contain harmful amounts of moisture at harvest-time, it becomes necessary to adjust the moisture to a safe level before storage. Various types of dryers for vegetable and herbage seeds have been described by North (1948). He places the maximum air temperature for safe drying at not more than 120°F., while some seeds, such as those of beans, onions, and leeks, should not be exposed to temperatures higher than 90°F. during drying. Christidis (1940), on the other hand, found that cotton seed can withstand the effects of drying at 60°C. for at least eleven hours, and at 90°C. for two to four minutes, without loss of vitality. He recommends 70° to 75°C. as an appropriate temperature range for drying these seeds. The time required (usually a few minutes to half an hour) to eliminate the excess moisture in cotton-seeds depends on the temperature and speed of the current of hot air used, as well as on the amount of moisture to be removed. Seeds of crimson clover (*Trifolium incarnatum*) can be cured, either by natural drying or by artificial heat of 43·3°C., to a moisture content of 10 per cent (Ward & Butt, 1955). Brewer & Butt (1950) dried seeds of the blue lupine (*Lupinus angustifolius* L.) at 115° to 140°F. (46° to 60°C.) in a forced-draught oven. Artificial drying of maize with heated air under forced-draught conditions is used extensively by growers of hybrid seed, according to Kiesselbach (1939). Under good management this practice can remove the danger of injury from freezing and also facilitate early harvest, processing, and storage. Kiesselbach recommended a reduction in moisture content of the seed to 12 to 13 per cent at a temperature range of 105° to 110°F. (40° to

43°C.), but when the initial moisture content is near 50 per cent, the drying temperature should be held to 105°F.

Sulphuric acid is a good drying agent, but has the disadvantage that fumes from the acid may injure the seeds. Also, it is impractical to use. Glycerine or calcium chloride, in either case mixed with water, have also been used (Nakajima, 1925). In addition to calcium chloride, Kondo (1926a) used lime, wood ash, and straw ash for mixing with seeds to control the moisture content.

Seeds may be dried by being placed over or mixed with such drying agents as calcium chloride, calcium oxide, or silica gel. Some specific effects of such drying methods will be discussed in the chapter on vegetable-seed storage. Germination as affected by storage over various hygroscopic substances has been described by Nakajima (1927).

For larger seed-lots, and especially for the seedsman who must maintain a high degree of viability in any surplus seed stocks, a dehumidified storage room is often useful in practice. The importance of the temperature in the storage chamber should not be overlooked, and this aspect will be discussed in the next chapter.

The advantage of pre-harvest drying of rice by chemicals sprayed from an airplane has been described (Anon., 1953a).

IV

OTHER FACTORS AFFECTING LONGEVITY

TEMPERATURE EFFECTS

JUST as it is impossible to discuss the effect of moisture content on the viability of seeds without consideration of the temperature, so it is impossible to list temperature effects without consideration of the moisture content. However, the data secured over a period of several years by many workers in the field show certain trends which are attributable to temperature.

In general, the higher the temperature the more rapid the deterioration at a given moisture level. Conversely, the lower the temperature the greater the tolerance of high moisture content—apart, of course, from the mechanical injury of the seed tissues at temperatures below freezing. Most often, experiments set up to show temperature effects have included only two or three temperatures; these were usually ordinary room temperature, which varied with the locality, and a lower temperature approximating to that of a cool cellar (5° to 10°C.). If the moisture content of the seed is not too high, i.e. if the humidity in the cold-storage room is low, maintenance at 5° to 10°C. will extend the life of the seeds far beyond that under similar humidity conditions at ordinary room temperature. A high humidity room at 5°C., however, will bring about much more rapid deterioration of seeds than a dry room at 25°C. This has been demonstrated repeatedly and will be discussed under methods of storage of the seeds of individual species.

Perhaps the most significant finding of recent years regarding the effect of temperature on seed viability is the discovery that below-freezing temperatures are superior to above-freezing ones for keeping several different kinds of seeds. Only a few instances will be cited here for illustration; others will be discussed under the individual species.

Conifer seed storage may serve as an example. The general literature on this subject will be reviewed in the chapter on tree seeds (Chapter IX). We are concerned here only with the evidence for the superiority of below-freezing temperatures for storage of these seeds. One of the first controlled experiments was that of Barton (1935a), who placed conifer seeds under different conditions at three temperatures: (1) room, (2) 5°C., and (3) in a refrigeration room with an average temperature of −15°C. for the first two or three years of the experiments, after which they were transferred to another refrigeration room with an average temperature of −5°C. The seeds used were of loblolly pine (*Pinus taeda* L.), slash pine (*Pinus caribaea* Morelet), shortleaf pine (*Pinus echinata* Mill.), longleaf pine (*Pinus palustris* Mill.), red pine (*Pinus resinosa* Ait.), *Pinus ponderosa* Douglas, *Picea excelsa* Link, and *Picea glauca* (Moench) Voss (*P. canadensis* of authors). The conclusion at the time of the first report of the results of these tests, when the seeds had been in storage up to seven years (Barton, 1935*a*), was that sealed storage at either of the low temperatures (5°C. or −5° to −15°C.) was effective for the maintenance of viability. However, the data showed a significant superiority of the lower of the two temperatures at that time (Wakeley, 1945). Further tests (Barton, 1953*a*) showed that some of these seeds remained viable for up to sixteen years, when the supply of seeds became exhausted. Isaac (1934) stored noble fir seed successfully for five years at 15°F. (−10°C.).

TABLE VII

GERMINATION AND MOISTURE CONTENT OF CONIFER SEEDS AFTER
STORAGE AT SUB-FREEZING TEMPERATURES

Species and per cent germination at time of storage	Approximate storage temp. (°C.)	Per cent germination after storage for 1, 2, 3, 4, and 5 yr.				
		1	2	3	4	5
Ponderosa pine	−4	87	91	92	90	81
95	−11	85	79	70	79	81
	−18	81	75	72	77	74
Douglas fir	−4	69	44	7	0	—
88	−11	85	75	60	33	22
	−18	88	80	78	61	65
Sitka spruce	−4	49	11	1	0	—
56	−11	62	38	25	8	2
	−18	60	55	47	43	50
Western red cedar	−4	66	27	9	1	0
60	−11	65	31	24	14	0
	−18	74	74	76	65	70
Western hemlock	−4	5	0	0	0	—
12	−11	7	0	0	0	—
	−18	19	13	13	9	9

— Indicates that no test was made.

With this knowledge, and because of a need for determining the best possible storage condition for valuable, short-lived conifer seeds, a test was started in which three different sub-freezing temperatures were used for storage of seeds of Ponderosa pine (*Pinus ponderosa* Dougl.), Douglas fir (*Pseudotsuga taxifolia* Britt.), Sitka spruce (*Picea sitchensis* Carr.), Western red cedar (*Thuja plicata* Donn.), and Western hemlock (*Tsuga heterophylla* Sarg.) (Barton, 1954). The available seeds of each species were divided into three lots which were stored in canvas bags at temperatures of approximately $-4°$, $-11°$, and $-18°C$. For the first two temperatures, cold rooms were used with average temperatures as indicated but with a fluctuation of $\pm5°C$. A food freezer furnished the more constant temperature of $-18°C$.

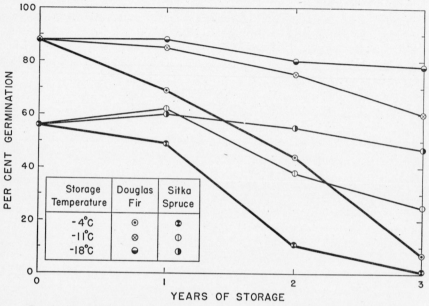

FIG. 6.—Graph indicating percentage germination of seeds of Douglas fir and Sitka spruce after storage at sub-freezing temperatures.

Viability tests were made at the beginning of the experiment and then annually for five years in succession. Results are shown in Table VII. There was great variation in the viability of the different species at the beginning of the experiment. The seeds of Ponderosa pine and Douglas fir were of excellent quality, giving 95 and 88 per cent germination, respectively, at the time storage was begun. The seeds of Sitka spruce and Western red cedar were less vigorous, with initial germination percentages of 56 and 60, while only 12 per cent of the Western hemlock seeds germinated.

All of the seeds, except those of Ponderosa pine, 'kept' better at $-18°C$. than at $-11°C$., and deterioration was most rapid at $-4°C$. (Table VII). It was anticipated that some such effect might be obtained, but the rapidity of the response was striking. Definite differences in deterioration rates were evident after only two years of storage. In experiments with other seeds (Barton, 1953*a*) at least five years of storage were required to show differences in response between 5° and −4°C. The germination capacity of Douglas fir and Sitka spruce seeds after storage at the three temperatures is shown graphically in Fig. 6. It appears that the lower the sub-freezing temperature (to −18°C.), the better the survival of Douglas fir and Sitka spruce seeds in canvas bags.

Other tests (Barton, 1954a) have shown that reduction in moisture content extends the life of Douglas fir and Western hemlock seeds even at the very favourable temperature of −18°C. Allen (1957) found 0°F. better than 32°F. for storage of conifer seeds in only a few cases out of twenty-seven lots representing eleven species. He stored the seeds in sealed vials for periods of from five to seven years.

Although these conifer seeds, together with maize and with vegetable and flower seeds (see chapters on these various seeds), have been benefited by storage below freezing, it should not be assumed that all seeds are so affected. Weibull (1955) points out that, while cold storage at −20°C. is good for maintaining the viability of many kinds of vegetable seed, it cannot be used for parsley (*Petroselinum sativum*) seed, which reacts adversely to cold storage and loses germination ability rapidly at −20°C. Also, the seeds of four types of flower, namely *Antirrhinum majus*, *Asparagus plumosus*, *Petunia hybrida*, and *Viola tricolor* var. *maxima*, are actually killed at −20°C. As a result of his experiments Weibull (1955, p. 667) concluded that 'a test should therefore be made before the large-scale application of cold storage to a certain kind of seed'. These results notwithstanding, the discovery of the successful storage of many types of seeds at sub-freezing temperatures eliminates the necessity for sealed storage and maintains a high quality in the seeds which are then able to resist further unfavourable storage conditions (see Chapter XI, on packeting, etc.).

It will be noted in Chapter VII that grains, especially maize, are now being stored at sub-freezing temperatures.

GAS EFFECTS

Gaseous exchange is important in the metabolism of the seed and hence has a relation to its life-span. Undoubtedly respiration is affected, but the factors involved in seed mortality appear to be much more complex than those in ordinary aerobic respiration. In any case, the gaseous exchange is directly related to the moisture content of the seed and to the temperature at which it is stored. Because moisture and temperature conditions seem to be more directly concerned in determining the life-span of seeds, less experimental work has been done on the effects of different gases.

A detailed experiment on cotton-seed storage in various gases under controlled temperature and moisture conditions has been reported by Simpson (1953). Initially, moisture content of the seeds used in this experiment was 10 per cent and the germination was 97 per cent. Moisture was adjusted to two levels, 7 per cent by drying in the sun and 13 per cent by adding water. The seeds were then placed in pint milk-bottles fitted with tight cork stoppers. The corks were pressed down below the rim of the bottle, leaving room for a paraffin seal. The gases used were air, oxygen, carbon dioxide, and nitrogen. The last three gases were introduced into the bottles by evacuating the latter and then flushing them with the desired gas twice before filling, adjusting the pressure to that of the atmosphere, and sealing with paraffin. Each bottle contained approximately 100 gm. of seed. The total volume of gas in each bottle was about 355 ml.

Examination of the samples was made every six months for three-and-a-half years. After that time, lots with 7 per cent moisture and at 70°F. were held in reserve and tested after eight and ten years of storage. Gas pressure and weight of free carbon dioxide within the bottle, and moisture content and germination percentages of the seeds, were obtained from duplicate samples on each sampling date. Chemical analyses were also made on seeds stored for two-and-a-half, three-and-a-half, and ten years.

Variation in moisture content during the storage period did not exceed that which would be expected from sampling error. Seeds containing 7 per cent of moisture were still fully viable after storage for ten years at 70°F., and there was no deterioration of seeds with this moisture content at 90°F. after three-and-a-half years, when the supply was exhausted. At the higher moisture content of 13 per cent, viability was retained for six months at 70°F., but deterioration was rapid after that time, and only a few seeds were capable of germination after one-and-a-half years. At 90°F., seeds with 13 per cent moisture failed to survive for as long as six months.

There was a decrease in gas pressure in the bottles with seeds sealed in air, oxygen, or carbon dioxide—most logically explained, according to the author (Simpson, 1953), by assuming that carbon dioxide was absorbed by the seed. Containers with atmospheres of nitrogen gas showed little change in pressure with storage condition or time. The carbon dioxide recovered from the containers after intervals of storage indicated rapid production of carbon dioxide by the seeds in an atmosphere of oxygen, moderate production in air, and low production in nitrogen. Considerably less carbon dioxide was recovered after storage than was placed in the containers at the beginning of the test, i.e. the seeds must have absorbed carbon dioxide. The author concluded that, within the limits of this experiment, the initial atmosphere in the storage chamber had no effect on the longevity of the cotton-seed, and that free oxygen was not essential to the continuous life-processes of the seed. Therefore, moisture and temperature of storage are more important than atmosphere in keeping cotton-seed for ten years.

Sayre (1940) sealed maize grain for five years in air, oxygen, carbon dioxide, and nitrogen. There was no deterioration in samples stored at

3°C. and −25°C. At 30°C., however, deterioration was evident in all samples after storage for three years, but loss of germination capacity was more pronounced in oxygen and nitrogen atmospheres than in air or carbon dioxide. As a result of these tests, the author questioned the theory that free oxygen is necessary for continued vitality of stored seed, and stated that respiration may speed deterioration and consequently any condition which will decrease respiration will extend the life of the seed. The reason for the harmful effect of the nitrogen atmosphere on stored maize grain is not clear.

Busse (1935) stored poplar seeds in air and in a partial vacuum and found that the life-span was increased by reducing the air pressure. He believed this to be due to the injurious action of the oxygen in the air. Guillaumin (1928) demonstrated by an experiment lasting twelve years that a vacuum was a means of preserving the viability of grains. An atmosphere devoid of oxygen was as good as a vacuum; hence the harmful effect of oxygen itself was shown. Sampietro (1931) found that nitrogen extended the life of rice seeds in storage, and that this gas was better than carbon dioxide, air, or a vacuum. Air and carbon dioxide had the same effect on rice in storage, according to Kondo & Okamura (1934).

Seeds of the American elm stored in an atmosphere of oxygen lost their viability more quickly than others sealed in air. When moisture or temperature conditions were unfavourable, a partial vacuum extended the life of these seeds (Barton, 1939a). Some benefits of a lessened oxygen supply may have been experienced by conifer seeds under certain storage conditions (Barton, 1935a), but aster and *Verbena* seeds were not improved by a partial vacuum under the conditions tried by Barton (1953a).

The fact that an atmosphere of pure oxygen increases the respiratory rate of seeds, leads one to expect that this gas might shorten the life of some types of seeds under certain conditions—especially those of un- favourable moisture and temperature—and perhaps more significantly for seeds of low germination capacity than for others. Further evidence of the harmful effect of oxygen has been demonstrated for barley seeds stored after X-radiation (Adams *et al.*, 1955). It was found that oxygen enhanced the amount of radiation damage as measured by frequency of chromosome bridges, rate of germination, and emergence of seedlings, while nitrogen retarded these effects.

It seems certain that most seeds can carry on the minimum metabolic activity necessary for retention of viability without an outside supply of oxygen, and hence would not be benefited by such a supply. Ohga (1926a) found, for example, that internal gas pumped out of very old but viable Indian lotus fruits was composed of an average of 18·33 per cent oxygen, 0·74 per cent carbon dioxide, and 80·93 per cent nitrogen—as compared to 18·88 per cent oxygen, 0·81 per cent carbon dioxide, and 80·31 per cent nitrogen in fresh fruits. The permeability of the seed-coat to gases is at a minimum in dry seeds and increases with moisture (Becquerel,

35

1904). This may well be a factor in the favourable effect of drying for many types of seeds. The whole matter of permeability is an important one, but is outside the scope of this book.

LIGHT EFFECTS

Although there are numerous publications concerning the effect of light on the germination of seeds, there is very little available information on the effect of light on seeds in storage. No effect on germination of exposure of eight different seed-types to the sun for forty-four days was noted by Tammes (1900). When the moisture content of seeds of *Cinchona ledgeriana* was as high as 9·4 per cent, there was, however, some evidence of a harmful effect of exposure to light during storage in the laboratory (Barton, 1947).

C. Jensen (1941) treated seeds of cauliflower by exposing them to a Quartz-lamp and a Sollux-lamp together, with full strength for one hour at a distance of 1 m. from the lamps. The seeds were stirred frequently during the exposure. Following this treatment the seeds were stored in glass containers plugged with an ordinary cork or a paraffined cork. Light-treated seeds maintained a higher germination percentage over a period of eight years than untreated seeds. Light treatment not only extended the life-span, but also increased the germination capacity and vigour of fresh seeds. No moisture determinations were made before or after treatment, and it is possible that the light-treatment procedure may have reduced the moisture content to a more favourable level. In any case, the whole subject of the effect of light during storage is a field which should be explored more thoroughly.

INHERITANCE

J. W. Jones (1926) noted that the seeds of some rice varieties appear to deteriorate with age more rapidly than others. Three-year-old seeds of five varieties ranged in germination from 85 to 99 per cent, while those of three other varieties of similar age germinated only from 19 to 72 per cent.

Maize seeds homozygous for $luteus_2$ and $luteus_4$ genes lose viability more quickly than seeds homozygous for any of the remaining six luteus genes (Weiss & Wentz, 1937). Fresh seeds with $luteus_2$ and $luteus_4$ genes germinated as rapidly as normal seeds, but such luteus seeds when a year old germinated much more slowly and the plants produced from them also grew more slowly than normal plants. The authors suggest that it is possible that these genes do not cause decrease in germinability, but that each of these two genes is closely linked with another gene which causes loss of viability.

Lindstrom (1942) also found that embryo longevity of maize is heritable. He made germination tests with hundreds of inbred lines and F_1 crosses of five- to twelve-year-old seeds stored at room temperature. They

36

showed a variation in germination of from 0 to 90 per cent. Further tests were made under controlled temperature and humidity storage conditions. It was found that hybrid seed of long-lived by short-lived parental inbreds exhibited the long-lived condition, often with significant differences in reciprocal crosses.

Although other examples of detailed experiments to determine the inheritance of longevity are lacking, it is obvious that the general make-up of the particular seed, especially the permeability of the coat, is a major factor in determining its life-span.

SEED-COAT EFFECTS

The importance of the seed-coat in the life of the seed was recognized early and has already been discussed (Chapter I). Rees (1911) gave evidence that the impermeable part of the coat was a waxy cuticle in many species, and the thicker the cuticle, the longer was the time in sulphuric acid required to produce swelling when treated seeds were placed in water. Wahlen (1929) found that longevity in clover seeds depended upon the impermeability of the coat.

Mechanical injury to seeds by threshing usually contributes to immediate reduction in germination capacity and to an accelerated loss of viability in storage. Certainly the embryos of such seeds are more vulnerable to attack by micro-organisms. Scarification of alfalfa seeds (Battle, 1948), mechanical dehulling of bromegrass seeds (Ahlgren et al., 1950), and delinting of cotton-seeds with sulphuric acid (Hamid, 1938), are all examples of treatments which reduce the life-span of the seeds. Leaving the chaff on stored seeds of several species and stocks of wheat and two varieties of barley, appears to be the best for storing ; but such seeds should be shelled before testing or planting, as removal of the chaff results in a considerable increase in the percentage of germination (L. Smith, 1948). On the other hand, Cutler (1940) found no indication of loss of viability attributable to rubbing or clipping oat grains.

Although the seeds of the longest known life-span are those which have remained impermeable during the storage period, it is sometimes important for planting purposes to have the seeds become permeable during storage. Conversely, seeds which are permeable when harvested may become impermeable during storage under certain conditions of temperature and humidity. While such 'hard' seeds are often still alive when tested, there are some in which a high percentage are not viable—e.g. *Vicia villosa* (V. K. Toole, 1939). There is a considerable amount of literature on the development of hard seeds in storage, but the subject will not be discussed further in this book.

MATURITY EFFECTS

Studies on the effect of maturity on the germination of seeds have been made principally to determine the proper date for harvesting of seed for sowing. The germination tests have been related to the performance of the seeds immediately after they had been harvested, rather than to their resistance to subsequent storage.

Seeds of Norway spruce and Scots pine ripen prior to the lignification of the cones, and hence cones harvested early will dry out more rapidly than those from a later harvest. Rapid drying accounts for the lower germination after a year of storage of seed from cones harvested early, according to G. Vincent & Freudl (1931).

To study the effect of maturity on the viability and longevity of the seeds of western range and pasture grasses, McAlister (1943) collected seeds of *Agropyron cristatum*, *A. smithii*, *A. trachycaulum*, *Bromus inermis*, *B. marginatus*, *B. polyanthus*, *Elymus glaucus*, and *Stipa viridula*, in the pre-milk, milk, dough, and mature stages of development. Greenhouse tests of the germination capacity of these seeds were made after storage for 4, 9, 15, 22, 40, 51, and 58 months. Pre-milk and milk seeds were inferior in germination and longevity to seeds harvested in either dough or mature stages. Exceptions to this were the pre-milk and milk stages of seeds of *Bromus marginatus* and *B. polyanthus*, which gave as high a percentage of germination as mature seeds during the whole period of storage. Dough-stage seed had similar viability and longevity to the mature seeds in all species, when tested in the greenhouse. In field plantings, however, the immature seeds were generally much inferior, as far as seedling emergence was concerned, to those harvested at maturity. The only immature seeds which gave as good a stand as the mature seeds during the three years following collection were those of the dough stage of *Bromus marginatus*. No differences in survival or size of plants from mature and immature seeds were evident by the end of the growing-season in the field.

Rice seed for storage and sowing should be saved from early and medium plantings as such grain has a lower moisture content at harvest time than has grain from late plantings (McNeal, 1950). It has been suggested by Riddell & Gries (1956) that the variations in growth of spring wheats from seed of different ages are related to the temperature during maturation rather than to the age of the seed or to conditions during storage.

DORMANCY EFFECTS

A knowledge of the presence of a dormant period is important in developing methods for testing seed viability. Dormancy as expressed by special requirements for germination makes possible the extended life of seeds buried in soil (*see* Chapter II). Dormancy imposed by an impermeable seed-coat is a significant factor in seed longevity, as we have already seen. The oldest known viable seeds are those with hard seed-coats which resist water absorption. Because of this character, seed-coats often cause delay in germination (Crocker, 1906; and others).

Embryo dormancy is also important in germination and hence in tests for viability, but is not directly related to longevity.

FRUIT OR PULP EFFECTS

Many studies have indicated the inhibiting effect of fruit or pulp on germination, but few authors have reported a subsequent effect on longevity.

Pulp disintegration of peaches in a pile of twenty bushels of fruit reduced the germination of the seeds because of the high temperature caused by rotting, and also because of some unknown property of the fermenting juice (Haut & Gardner, 1935). Germination tests of seeds removed from mature and immature fruits of butternut squash have been made after 211 days of storage (Holmes, 1953). All the seeds germinated well.

V

OTHER FACTORS AFFECTING LONGEVITY (contd.)

MICROFLORA

THE role of bacteria and fungi associated with seeds in storage in bringing about their deterioration has been a subject of interest for many years. This interest has led to much detailed work on the subject. Two excellent reviews (Semeniuk & Gilman, 1944; Christensen, 1957), and a chapter on microflora written by Semeniuk in a recent book on *Storage of Cereal Grains and their Products* (Anderson & Alcock, 1954), should be consulted by the reader for the present status of the problem, and for a complete review of the literature on the subject.

Workers at the University of Minnesota have published a series of twenty-five articles on grain storage studies from 1945 to 1957. The titles of some of these publications indicate special interest in the microflora of stored seeds, especially soya beans, wheat, and cotton-seed (Milner & Geddes, 1946, 1946a; Milner *et al.*, 1947, 1947a; Christensen *et al.*, 1949; Hummel *et al.*, 1954; Christensen, 1955, 1955a; Sorger-Domenigg *et al.*, 1955, 1955a; Papavizas & Christensen, 1957; Tuite & Christensen, 1957). Some of the more recent of these studies will now be discussed.

WHEAT

It is to be expected that microflora will grow better on seeds with high moisture contents, resulting eventually in excess heating and mould growth when seeds are stored under conditions of high humidity and temperatures favourable for fungus growth. Christensen (1955) investigated the invasion by moulds of wheat grain stored in the laboratory for sixteen months and containing 13·5 to 15.0 per cent of moisture. He found that *Aspergillus restrictus* invaded grains under these conditions and that the invasion was associated with decreasing germination and increasing development of a brown colour of the germ which is characteristic of 'sick' wheat. The main invasion by the fungus appeared to be in the region of the germ. These results indicate that a moisture content of 13·5 per cent, which has been commonly accepted as a safe one for grain storage, is too high, as it does permit slow invasion by this fungus.

In further studies, the effect of mould growth (during temporary exposure of wheat grain to conditions of high moisture) upon the development of germ damage during subsequent storage was determined (Sorger-Domenigg *et al.*, 1955). Seeds of 'hard' red spring wheat which had been inoculated with an appropriate volume of water suspension of mould spores to give samples containing 15, 18, and 21 per cent moisture. After five, ten or fifteen days incubation, the wheat samples were air-dried to approximately 13 or 14 per cent moisture content on trays at room temperature, where they were then stored. High levels of mould infestation increased the development of germ damage, and this was accompanied by loss in germination capacity and by increased 'fat acidity' during the temporary storage at high moisture levels. During subsequent storage at 13 and 14 per cent moisture, the mould population of the majority of the samples decreased, but the germ damage and fat acidity continued to increase, especially in the samples which had been heavily infested with mould. Not only did inoculated samples temporarily exposed to 21 per cent moisture show loss of viability, but seeds so treated gave flour of high ash content, poor colour, poor baking strength, and general loss in baking quality. Negative correlations were found between free fatty acids and loaf volume, germ damage and loaf volume, and the logarithm of viability and free fatty acids. It was pointed out that losses in viability preceded the discoloration of the germ and were indicative of incipient damage and poor storage properties. The authors state (Sorger-Domenigg et al., 1955, p. 282): 'While not excluding the possibility that sick wheat may develop in the absence of mold growth, the present investigations provide strong evidence that molds are largely responsible for the losses in viability which precede the discoloration of the germ and the increases in fat acidity.' However, it would appear that the sharp decrease in the mould count in wheat stored at 14 per cent or less moisture after exposure to high moisture, even though other tests such as viability, germ damage, and baking quality indicated continued decline of seed quality, might argue against the mould invasion as the sole cause of deterioration.

Papavizas & Christensen (1957) made a study of various fungi associated with wheat seeds in storage, to determine the effect of each fungus on the development of 'sick' wheat. They stored grains after sterilizing with sodium hypochlorite (1 per cent for one minute) and drying to a moisture content of 11 to 12 per cent. They point out that storage fungi do not invade seeds to any appreciable extent before harvest, and that wheat seeds sterilized in this way were almost completely free from fungi. Some of the seeds were then inoculated with dry spores of *Aspergillus candidus*, *A. repens*, or *A. restrictus*, and stored for up to seven months in desiccators at 25°C. over humidifying solutions giving moisture contents of from 14·7 to approximately 20 per cent. Non-inoculated seeds declined very little in germination capacity and produced very few damaged germs ('sick' wheat), while those inoculated with *Aspergillus* deteriorated much more rapidly. This is illustrated in Fig. 7 taken from the original article (Papavizas & Christensen, 1957). Also, of the three species of *Aspergillus* indicated in the figure, *A. repens* was least harmful to the stored seeds. Invasion of the wheat seed by fungi was followed by

41

the development of germ damage or 'sick' wheat as pictured in Plate 2 (from Papavizas & Christensen, 1957).

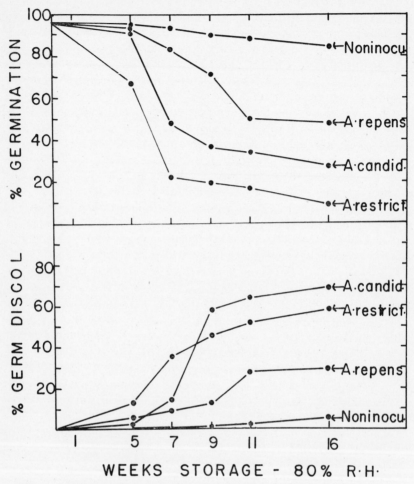

FIG. 7.—Graphs showing germination and germ damage of grains of Willet wheat non-inoculated and inoculated with *Aspergillus candidus*, *A. repens*, and *A. restrictus*, and stored at 25°C. under 80 per cent relative humidity (moisture content 16·0 to 16·4 per cent) (from Papavizas & Christensen, 1957; courtesy of Department of Plant Pathology and Botany, University of Minnesota, St. Paul, Minnesota, U.S.A.).

Fusarium moniliforme appeared to bring about more rapid deterioration of stored seeds of barley than did species of *Aspergillus* and *Penicillium* (Armolik *et al.*, 1956). These studies were carried further by treating barley seeds with twenty-day-old culture filtrates of several fungi. All of the filtrates, except those from *Penicillium chrysogenum*, reduced early seedling growth (*see* Plate 3). *Fusarium moniliforme*, which caused the worst damage to seeds in storage, also had the most deleterious effect in filtrate form.

Although Armolik *et al.* (1956) concluded that the fungi were important factors in the deterioration of moist barley seeds, they also noted deterioration of the sterilized seeds, thus indicating that fungi are not the only factors concerned. Further tests on the humidity requirements for germination of the spores of fungi associated with grain, showed that they would develop on substrates at moisture contents of between 10 and 20 per cent, depending upon the fungus (Armolik & Dickson, 1956). Members of the *Aspergillus glaucus* group germinated under lower humidity conditions than did the others, and they were usually the first fungi to invade cereal grains in storage. *Fusarium moniliforme*, on the other hand, required 20·4 per cent of moisture for germination, which was the highest of any fungus studied. The moisture content of freshly 'combined' barley grain was found to be high enough to support growth of all of the fungi studied—which indicates the need for artificial drying, as deterioration is very rapid during the first ten days after harvesting.

To determine the time of invasion of the wheat seeds by forms of *Aspergillus* responsible for deterioration in storage, Tuite & Christensen (1957) collected grains from different varieties and classes of wheat from several different states over a period of three harvest seasons, and determined the presence of storage fungi by culturing the seeds on malt-salt agar. Only a small percentage of the seeds were found to be infected. However, storage fungi invaded the seeds readily when they were stored in a humid chamber, the threshed seeds being more susceptible than seeds left in the heads. The percentage of wheat seeds yielding storage fungi increased between harvest and the arrival of the grain at the storage site.

Further evidence of the causal relationship of mould to seed deterioration was furnished by Christensen (1955*a*) in his study of the incidence of germ damage in commercial wheat grain. Christensen's abstract of his findings is as follows:

Germination of the seed, and number and kinds of molds present, were determined in 'sick' and sound seeds picked from 26 commercial samples containing from 5–55 per cent 'sick' wheat, and in sound wheat from bulks in which no deterioration had occurred. The germination of 'sick' seed always was zero; molds were microscopically visible on the germs of 49 per cent of the seeds; they had an average mold count of 402,000/g.; and 94 per cent of the surface-disinfected seeds yielded storage molds. The 'sound' seeds picked from the lots in which deterioration had occurred had an average germination of 43 per cent and an average mold count of 32,000/g., and 84 per cent of the surface-disinfected seeds yielded storage molds. The really sound seeds from bulks in which no deterioration had occurred had an average germination of 91 per cent, an average mold count of less than 1,000/g., and 27 per cent of the surface-disinfected seeds yielded storage molds. The major fungi present in the 'sick' seeds were *Aspergillus restrictus*, *A. repens*, *A. candidus*, and *A. flavus*. Judged by various microscopic and cultural technics, all samples of 'sick' wheat had been very heavily invaded by storage molds; all of the evidence indicated that invasion of the germs of the seeds by these molds had preceded decrease in germination and increase in 'sick' wheat. In commercial storage, it seems very probable that invasion of the germs of the seed by common species of *Aspergillus* is a common cause of 'sick' wheat.

43

Some measurements have been made of the chemical changes taking place in mould-infected as compared with uninfected wheat grain. Increased respiratory rate, increased fat acidity, and decreased non-reducing sugars, were associated with the presence of mould (Hummel *et al.*, 1954). Loss in viability accompanied these changes, though at moisture levels of 18 per cent or above none of the seeds survived.

Cole & Milner (1953) presented evidence that the discoloration of the germs of 'sick' wheat grain is due to a browning reaction of the Maillard type. They found a positive correlation of 0·748 between the fluorescence of aqueous extracts of the sample and an evaluation of germ damage by federal inspectors. They also found a significant negative correlation of 0·775 between fluorescence and germination capacity of the samples. They suggested that the fluorescence test could be standardized as an index of viability.

Sorger-Domenigg *et al.* (1955) did not find such high correlation coefficients between fluorescence value and germ damage or germination capacity. The correlation between germ damage and formazan value (*see* Chapter XIII), though somewhat higher, still did not justify its use to predict the value of wheat seed. However, there was a high correlation (−0.92) between viability and fat acidity. B. Thomas (1941) pointed out that destruction by bacteria can occur in cereals with moisture contents of over 17 per cent.

OTHER SEEDS

Although the presence of many fungi, among which species of *Penicillium*, *Aspergillus*, and *Rhizopus* are predominant, has been demonstrated on stored maize grains, it is not thought to be of significance in the storage problem—chiefly because of the high moisture requirements of these fungi. Maize is usually dried to a moisture content below that required for the growth of the fungi (approximately 12 per cent) before it is stored under conditions which prevent atmospheric condensation (Semeniuk & Gilman, 1944; Semeniuk *et al.*, 1947).

Experiments of Venkatram (1950) showed that fungi were not the primary cause of cotton-seed deterioration, which was due rather to the high moisture content of the seeds and to unfavourable temperatures during storage.

Kurata *et al.* (1957) found that micro-organisms, including fungi, bacteria, yeasts, and actinomycetes, were not normally present in milled rice produced commercially in the southern area of the United States in 1954.

Whether fungi play a major role in the deterioration of seeds in storage still remains a question in the minds of many seed experts. The fungi grow only under high humidity conditions which favour deterioration of seeds not obviously invaded by fungi. Furthermore, although low temperature and low moisture content prolong the life of many kinds of seeds, the embryos finally do die under these conditions—without mould growth, which is inhibited by the storage conditions. Also, it has been shown above that sterile wheat seeds deteriorate under conditions in which mould invasion accelerates the deterioration. It becomes a question of whether the mould does not attack the deteriorating seeds. This theory is strengthened by the fact that most of the fungi found on stored seeds are saprophytes. It is an invariable rule that seeds of high quality can be germinated on a moist substrate without infection of the culture by bacteria and moulds. Germination is rapid and the seeds and seedlings remain apparently clean. However, seeds of the same lot which have been in sealed containers under conditions that cause deterioration but do not permit fungus growth, will rapidly become mouldy when they are placed on the germination substrate. Undoubtedly, both seeds harboured the same microflora at the beginning of the storage period. Whether the sound seeds produce a chemical deterrent to mould growth, or whether sugars or other fungus nutrients leach out from deteriorating seeds, is a matter for future tests to determine.

In cold tests used to evaluate some of the factors contributing to the longevity of maize seeds, it has been noted by Goodsell et al. (1955) that there is an increasing susceptibility to soil-borne pathogens with an increase in age of the seeds. They note that this phenomenon is not well understood but may perhaps be related to permeability changes.

A knowledge of biochemical changes which take place in deteriorating seeds under dry conditions at low temperatures over long periods of time, as compared to those taking place under moist conditions in a short time, would aid greatly in the interpretation of the role of microflora in the loss of viability by seeds.

LIFE-SPAN OF FUNGI

The life-span of the fungi associated with seeds in storage is a matter of some interest. It is not within the scope of this book to discuss longevity of fungi in detail, but a few examples will be given. The longevity of two fungi occurring on wheat seeds was compared by Russell (1956), who found that whereas *Alternaria tenuis* disappeared after seven years of storage, 8 per cent of the *Helminthosporium sativum* was still alive after seventeen years. The seeds themselves retained their full viability for about nine years under the conditions tried by Russell, thus indicating that prolonged storage of the seeds could be used to eliminate the *Alternaria* but not the *Helminthosporium*.

Machacek & Wallace (1952) stored wheat, oat, and barley seeds, infected with fungi, in paper bags in galvanized metal boxes at room temperature. At intervals up to ten years later, samples of the seeds were surface-sterilized and tested for the presence of the fungi. *Alternaria tenuis* (*sensu* Wilshire) in all three types of seeds, *Helminthosporium sativum* P.K. & B. in wheat and barley grains, and *Septoria nodorum* Berk.

in wheat grains, all deteriorated rapidly after one year of storage. *Helminthosporium avenae* Eidam in oats and *H. teres* Sacc. in barley, deteriorated more slowly, but a small percentage of each of the fungi was recovered after seven to ten years of storage. The pathogens remaining alive at the end of ten years when the seeds were still viable were considered to be of small consequence as far as crop production was concerned.

E. Brown & Robert (1943) found that *Alternaria* sp. on oats, wheat, and barley seeds were killed when stored for from six to fifteen months at 105°F., although the seeds would still germinate. This treatment should be used with caution, however, as seeds so treated may not be vigorous enough to produce a good crop, and prolonged storage at such a high temperature would certainly result in their death.

In rice seeds containing 18 to 20 per cent of moisture which had been sealed for seven months, some moulds survived but did not increase, aerobic bacteria remained the same or decreased in number, while facultative anaerobic organisms and yeasts increased in number (Teunisson, 1954). There was a loss in viability of the rice seeds during this period. Ageing of maize seeds reduced the prevalence of viable fungi in infected seed grain (Dungan & Koehler, 1944). During the first year of storage, *Nigrospora oryzae* infection was greatly reduced, *Diplodia zeae* was reduced about 50 per cent, and *Gibberella zeae* was almost completely eliminated. Although *Cephalosporium acremonium* and *Fusarium moniliforme* infections were reduced in the stored maize grain after three years, traces of them remained after up to seven and eight years of storage. Filutowicz & Bejnar (1954) observed no relation between the age of sugar beet seed and the degree of infection by *Cercospora beticola*.

CHEMICAL EFFECTS

The importance of micro-organisms in the safe storage of seeds has been demonstrated. Prevention of the destruction of the seeds by either fungi or insects can be controlled by chemical treatment of the seeds. Many new materials for such uses have become available within the last twenty-five years. For the most part, however, seeds have been treated immediately before planting. For long-term storage of chemically treated seeds there remains the question of the effect of the chemical on longevity. A considerable amount of work has already been done along this line.

The use of recommended methods of treating seeds with organic mercury compounds is greatly restricted when the seeds are to be kept for some years. Clayton (1931) found that treatment of vegetable seeds with mercuric chloride, liquid organic mercurials, or hot water, greatly shortened the life of vegetable seeds—even when it caused no immediate, apparent injury. Dust treatments were safer. From a case of injury to wheat germination caused by treatment with Ceresan, it was concluded that under improper storage conditions treated seeds may be permanently damaged (Crosier, 1930–3). Initial seed injury caused by careless handling is also a factor in damage by chemicals. Some fungicidal dusts, especially those containing ethyl mercury phosphate, have been found to depress germination of wheat grain after storage at 20°C. or higher (Wark, 1942; Koehler & Bever, 1956). An organo-mercury dressing of wheat seeds resulted in an increased percentage of abnormal seedlings when the seeds dusted had been superficially dry, the percentage of abnormality rising to as much as 15·3 when the seed had been superficially moist at the time of treatment (Brett & Weston, 1941). Storage of treated wheat seeds in closed containers caused rapid loss of viability, while untreated seeds under the same storage conditions retained a high germination capacity for more than a year. Over-treatment of cereal grains with the organic mercury compound, Panogen, did not appreciably impair the germination after eight months of storage; but longer storage periods might cause the development of injury from large dosages (Andrén, 1945). Fumigation of wheat grain with various chemicals is apt to result in reduction or loss of germination capacity, but does not necessarily impair baking quality (Cotton et al., 1946).

The effect of seven different mercurials on maize seeds in storage was studied by Koehler (1938). Some injury resulted, but when the storage was in a warm building or in a closed, unheated warehouse, the yields of plants grown from stored treated seed were frequently as good as or even better than those from freshly harvested seed. Treated maize seed was definitely injured by storage under the roof of an open shed.

Kreitlow & Garber (1946) stored seeds of alfalfa, red clover, Ladino clover, and Sudan grass, after treatment with New Improved Ceresan, Semesan, Arasan, Spergon, and Yellow Cuprocide. Open and sealed containers were used at 10° and 25°C. Germination tests made periodically in Petri dishes and in soil showed no injury to alfalfa, red clover, and Ladino clover, from any of the fungicides. Sudan grass seeds were injured by treatment with New Improved Ceresan, as shown by reduced germination within a month from treatment and by their complete loss of viability within six months. None of the other chemicals used reduced Sudan grass germination.

Ceresan treatment, wet or dry, had no effect on the germinability of flax-seed stored at moderate temperatures and low humidity for three-and-a-half years in New Zealand (Black, 1946).

Proper drying for at least four days, followed by treatment with organo-mercury compounds, has been recommended as a treatment for jute seeds (Ghosh et al., 1951).

Polyploidy was induced in stored beet seeds by treatment with Improved Ceresan (Lynes, 1945). After one year of storage the induced polyploidy ranged from 0 to 17·30 per cent, and seeds stored for two years showed from 1·13 to 69·02 per cent induction. The occurrence of polyploids in the field would be of little importance, for few of them would survive. However, a large percentage of polyploids could mean a much reduced stand.

Cotton-seed may be treated with organic mercury dusts at any time after harvest and thereafter stored for periods of up to seventeen months without injurious effects resulting from the treatment, and without decreasing the beneficial effects of the treatment in increasing seedling emergence and yield (Miles, 1939, 1941).

Sulpha drugs have been used to inhibit the mould respiration in damp grain (Milner, 1946). Finely ground sulphonamides, in concentrations of from 1:2,000 to 1:250 (ratio of weight of sulphanilamide to weight of damp seeds of wheat), were dusted onto the seeds. The strongest concentration of the chemical, 1:250, prevented increased respiration and resulted in 84 per cent germination as compared with 10 per cent in the controls.

Five commercial brands of fire extinguishers applied at a dosage rate of 0·5 oz. (14 gm.) per lb. on six seed stocks (*Lactuca sativa*, *Lolium* sp., *Triticum aestivum*, *Hordeum vulgare*, *Medicago sativa*, and *Beta vulgaris*) did not reduce viability (Suskin, 1953a). This is important, for fire in a seed warehouse, or water used to quench the fire, can cause complete loss of valuable seed stocks.

Studies made at the Southern Regional Research Laboratory, New Orleans, Louisiana, over a period of several years, have been directed towards a chemical spray treatment of cotton-seed which would reduce the increased respiration in storage and hence the resulting production of heat which causes deterioration. A number of chemicals were found to be promising (Lambou *et al.*, 1948) when used on seeds containing 12 per cent of moisture. Among these were propylene glycol dipropionate and 4,6-bis-chloromethyl xylene. Combinations of these two chemicals were also effective in maintaining viability for more than six months and for reducing the free fatty acids formed in the seeds during the storage period. Seeds were sprayed with the mixtures as they were moved by mechanical conveyor to large storage bins. Condon *et al.* (1949) found that ethylene chlorhydrin and related compounds inhibited heating in moist flax- and cotton-seeds and decreased fatty acid formation.

In addition to those reported by Lambou *et al.* (1948) and Condon *et al.* (1949), some 300 other chemicals have been tested at the Southern Regional Research Laboratory, and about forty-five of these, when used in low concentrations, have been found to reduce the amount of heating caused by decay in cotton-seed (Altschul, 1949). Contrary to other seed treatments which are designed to control organisms associated with the seed and its deterioration, these chemicals have their effect on the seed itself, through its metabolism. This method has been patented (Altschul *et al.*, 1951).

Control of infestation by insects during storage, especially of grain seeds, is of great economic importance. Besides the numerous insecticides now on the market, naphthalene has been used to control insects on

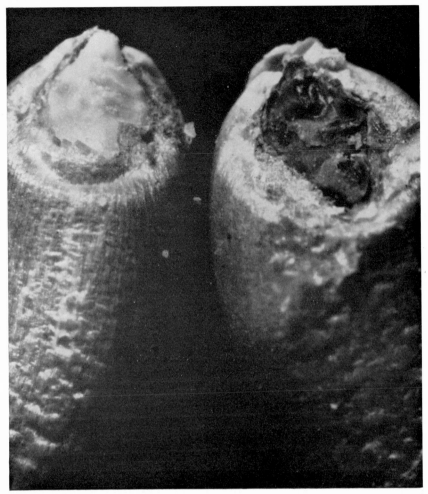

Photo. by courtesy Department of Plant Pathology and Botany, University of Minnesota, St Paul, Minnesota, U.S.A.

PLATE 2

Sound and germ-damaged wheat seed. Both were stored at 75 per cent relative humidity (moisture content 14·7 to 14·9 per cent) and at 25°C. for seven months. Left, non-inoculated remained sound. Right, inoculated with *Aspergillus restrictus* became germ-damaged. (From Papavizas & Christensen, 1957.)

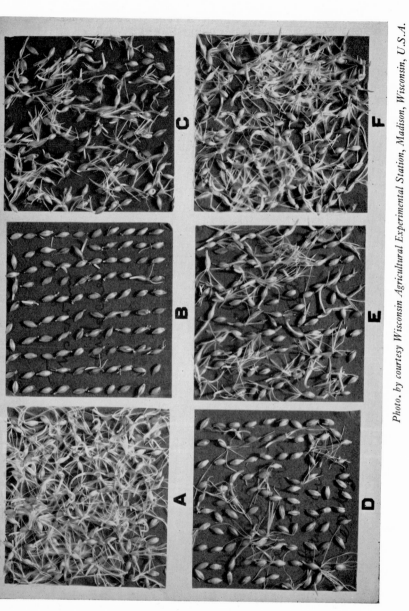

PLATE 3

The effect on seed germination and seedling development of twenty-days-old culture filtrates of several Fungi. Barley kernels were steeped for twenty-four hours in autoclaved filtrates. A, Control (Richard's Medium). B, *Fusarium moniliforme*. C, *Aspergillus flavus*. D, *A. niger*. E, *Penicillium cyclopium*. F, *P. chrysogenum*. (From Armolik, Dickson & Dickson, 1956.)

vegetable seeds (S. L. Singh & Singh, 1935). No effort will be made here to review the literature on insect control. Recent articles by Cotton *et al.* (1953), and S. W. Bailey (1955, 1956) deal with this subject. The physical conditions of seed storage, such as below-freezing temperatures and desiccation, may prevent insect infestation or destroy the insects already present. The rice weevil, for example, cannot breed in a seed with a moisture content of 8 per cent or less, and will soon die if restricted to such seeds for food (Cotton & Frankenfeld, 1945). Other insects, such as the flour beetles, on the contrary, are capable of breeding in very dry seeds, but can be controlled by low-temperature storage of the seeds. Two interesting methods for the detection of invisible insects in seeds are (i) carbon dioxide output of the grain samples (Howe & Oxley, 1944), and (ii) radiographs of the seeds which disclose the insects inside (Milner *et al.*, 1952).

VI

LONGEVITY OF VEGETABLE AND FLOWER SEEDS

THE next three chapters will deal with the viability of seeds of specific types under various conditions.

VEGETABLE SEEDS

The storage of vegetable seeds in such a way that seedling vigour as well as viability may be retained over a period of years, is a vital problem to both seedsmen and growers. As we have seen, successful storage conditions must include major consideration of temperature, moisture content, and oxygen supply. The effectiveness of the method depends on the interrelation of these factors rather than on a single factor.

One of the earliest reports on the subject was that of Kunzé (1882), who listed types of herb and vegetable seeds divided into life-spans of from one year to eight to ten years. Those with the shortest life-span were *Coriandrum sativum*, *Solanum melongena*, and *Satureja hortensis*; those which lived longest were *Cucumis sativus*, *Cucumis melo*, and *Cucurbita pepo*. Carruthers (1911) stored forty-three kinds of farm seeds in paper bags in cabinet drawers. His tests showed that the seeds varied in the rapidity of the decline in germination. Carrot seeds lost their vitality at a uniform rate and gave no germination in the tenth year.

The storage of vegetable seeds under controlled conditions is comparatively recent. Joseph (1929a), working with parsnip seeds, cited the beneficial effect of low temperature and desiccation on keeping quality, and Beattie & Tatman (1950) kept parsnip seeds at low temperatures for six years. Parsnip seeds lose viability rapidly when subjected to changes of temperature and humidity (Myers, 1935). Rose (1915), in his report of a study of delayed germination in economically important seeds, remarked that two varieties of lettuce seed improved in viability as they grew older, at least up to the end of the fourth year. This improvement, he said, was due to the increased permeability of the inner seed-coat. In his progress report on various kinds of seeds harvested in 1924 and stored through 1932, Pritchard (1933) listed those of carrot, cauliflower, lettuce, and onion as seeds which lost their germinative ability to a great extent, while seeds of tomato were among those retaining their viability very well.

A preliminary report on the effects of different storage conditions on the retention of viability of some vegetable seeds was published by Barton

(1935). Seeds of carrot (*Daucus carota* L. var. *sativa* DC.), cauliflower (*Brassica oleracea* L. var. *botrytis* L.), eggplant 'aubergine' (*Solanum melongena* L.), lettuce (*Lactuca sativa* L.), onion (*Allium cepa* L.), pepper (*Capsicum frutescens* L. [*C. annuum* L.]), and tomato (*Lycopersicum esculentum* Mill.) were stored, in 1932, at room temperature and at approximately $-4°$C. Samples of each kind of seed, with the original moisture content, were stored both open and sealed. In other samples part of the moisture was removed, either by drying in a desiccator over calcium oxide until about one-third of the moisture was removed, or by mixing directly with the amount of calcium oxide required to remove one-third or one-half of the moisture present in the seeds. All of the dried lots were sealed and stored at the two temperatures mentioned above. Sealed storage was in tin cans with tight-fitting lids, sealed for the first few years with Dekhotinsky cement and later with sealing wax. Care was taken not to heat the seeds in sealing.

Tests for viability were conducted both in electrically-controlled ovens and in the greenhouse. The oven tests were made on moist filter-paper in Petri dishes, while the greenhouse tests were made in soil composed of peat, sand, and sod soil, mixed in equal parts. In the Petri dishes the appearance of the hypocotyl was taken as the criterion of germination, while, in the greenhouse, actual seedling production, or the emergence of the shoot above ground, was recorded. This distinction is especially necessary with seeds that have been stored for some time, as old seeds often possess enough energy to send out a short root but cannot produce seedlings when planted in soil. Lots of 100 seeds each were used for the germination tests. Preliminary trials at different temperatures indicated the following to be best for oven tests of the various seeds: carrot, $15°$ to $30°$C. daily alternation; cauliflower, $25°$C. constant; eggplant 'aubergine', $20°$ to $30°$C. daily alternation; lettuce, $20°$C. constant or $15°$ to $30°$C. daily alternation; onion, $25°$C. constant; pepper, $25°$C. constant; tomato, $25°$C. constant.

Germination tests have been made on the surviving seeds each year since their storage. Selected data obtained after up to twenty years of storage are shown in Table VIII. Other records are to be found in the original publications (Barton, 1935, 1939, 1953a). Periods of up to six years demonstrated tomato and eggplant seeds as the most resistant to laboratory temperature-regime storage. Lettuce and onion seeds were the least resistant, and carrot and pepper seeds were intermediate, in their response to this temperature regime. Further tests revealed that carrot, eggplant 'aubergine', and tomato seeds, were still capable of about 50 per cent germination after open storage for eight years in the laboratory (Table VIII). The life-span was extended by drying the seeds before sealing—either by placing the seed samples over calcium oxide in a desiccator or by mixing the seeds directly with weighed amounts of calcium oxide, after which the lots were placed in sealed containers.

51

TABLE VIII

GERMINATION OF VEGETABLE SEEDS ON MOIST FILTER-PAPER AT CONTROLLED TEMPERATURES AFTER STORAGE UNDER VARIOUS CONDITIONS

Seed	Storage Container	Storage Moisture content per cent†	Per cent germination after storage in laboratory for years indicated							Per cent germination after storage at −4°C. for years indicated						
			1	2	4	8	12	16	20	1	2	4	8	12	16	20
Carrot 67*	Open	Air-dry	66	62	54	46	2	—	—	68	69	63	66	62	64	63
	Sealed	10·7	60	60	0	—	—	—	—	67	71	64	69	68	74	65
		a	62	65	55	62	56	43	2	67	71	72	74	69	68	63
		b	59	65	53	57	54	35	1	65	67	63	71	65	73	65
		c	63	72	62	65	52	48	6	69	71	62	69	66	66	69
Eggplant 86*	Open	Air-dry	81	90	73	59	17	—	—	82	85	80	79	—	—	—
	Sealed	10·4	86	74	68	37	0	—	—	85	87	79	86	84	—	—
		a	87	90	70	75	80	—	—	86	90	86	83	76	—	—
		b	87	86	69	80	66	54	10	79	82	80	84	83	82	86
		c	93	88	68	78	84	77	42	82	83	81	87	83	88	86
Lettuce 98*	Open	Air-dry	95	89	0	—	—	—	—	94	94	22	94	93	95	88
	Sealed	8·2	92	91	0	—	—	—	—	95	96	47	96	87	81	86
		a	92	97	65	69	67	—	—	91	94	75	93	80	77	80
		b	94	93	52	66	1	—	—	94	93	38	93	83	82	76
		c	94	89	26	84	3	0	—	91	95	48	95	88	85	84
Onion 98*	Open	Air-dry	62	69	0	—	—	—	—	94	96	87	85	—	—	—
	Sealed	12·5	82	1	0	—	—	—	—	96	94	92	85	—	—	—
		a	90	96	82	62	—	—	—	94	94	94	94	—	—	—
		b	96	93	63	33	1	—	—	95	93	94	96	89	88	—
		c	96	95	91	76	32	—	—	97	96	87	96	95	87	93
Pepper 73*	Open	Air-dry	67	64	16	—	—	—	—	80	86	66	69	62	57	19
	Sealed	10·4	22	0	0	—	—	—	—	75	86	72	72	73	64	—
		a	74	74	32	1	0	—	—	70	81	73	74	61	72	54
		b	74	74	28	4	0	—	—	76	73	73	66	71	70	67
		c	76	74	59	39	1	—	—	43	73	62	63	70	62	61
Tomato 93*	Open	Air-dry	94	89	75	48	3	—	—	92	90	78	86	88	—	—
	Sealed	10·0	91	84	46	0	—	—	—	93	89	91	91	—	—	—
		a	87	87	85	80	—	—	—	92	87	90	91	—	—	—
		b	91	92	91	90	75	45	1	90	90	89	92	91	90	—
		c	90	89	90	86	79	71	23	90	91	83	90	92	92	88

* Per cent germination at beginning of storage.

† At the beginning of the storage period.

a = Dried over CaO to remove about one-third of indicated moisture; b = mixed with sufficient CaO to remove about one-third of the indicated moisture; c = mixed with sufficient CaO to remove about one-half of the indicated moisture.

— Indicates that no test was made.

Although some vegetable seeds remained viable for several years at laboratory temperatures, -4°C. proved far superior for keeping the seeds (Table VIII). Even onion seeds, the germination capacity of which is seriously impaired after only a single year in open storage in the laboratory, gave 93 per cent germination twenty years after they had been mixed with calcium oxide, placed in sealed containers, and kept at -4°C. continuously.

Some vegetable seeds were stored by Pedro (1936) with and without calcium chloride at temperatures ranging from 0° to 28°C. The beneficial effect of drying was shown significantly by all seeds after 312 days in storage—with the exception of tomato seeds which 'kept' well in all conditions. From his results the author concluded that dry, warm storage was better than moist, cool storage. Hence he considered moisture to be more detrimental than temperature. Vegetable seeds have been kept in Japan for ten years in air-tight sealed containers with calcium chloride (Kondo & Kasahara, 1941).

Because onion seeds lose their viability rapidly, they have been the subject of special investigations (Beattie & Boswell, 1939; Brison, 1942; F. R. Brown, 1942; Myers, 1942; Nutile, 1957). All of these papers report the advantage of drying, sealing, and low temperature, for maintaining the viability of these seeds, and Brison (1942) pointed out that in room temperature storage for forty-four months, onion seeds sealed under a vacuum proved slightly better than those sealed in air, though this was not apparent in cold storage.

The effects of moisture and temperature on the keeping-quality of onion seeds has been studied by Beattie & Boswell (1938), who reported that at both 40°F. and room temperature-regime storage, 10 per cent of moisture resulted in greater loss of germination than 6 or 8 per cent, but was without effect at 20°F. $(-7^\circ$C.). Seeds of high moisture contents kept at 40°F. or room temperatures showed better germination after open than after sealed storage.

Many investigators have stored vegetable seeds under a variety of conditions and have reported the germination behaviour of these seeds after storage. Among such authors are Pritchard (1933), Stone (1938), Boswell et al. (1940), Rodrigo & Tecson (1940), E. H. Toole (1942), E. H. Toole et al. (1948), Harrison (1956), and Madsen (1957).

In addition to the literature dealing with vegetable seeds in general, there are numerous reports on the keeping quality of specific types.

A comprehensive physiological and microchemical study of lettuce seed viability has been made by Griffiths (1942). After two years of storage, deterioration of tomato seeds was evident from 'natural' storage temperatures, but no deterioration was evident after storage at approximately 40°F. (Myers, 1946). Beans of various kinds were the subject of investigations by Garner & Sanders (1935), Rodrigo (1939), E. H. Toole & Toole (1953), and Dexter et al. (1955). Data on radish (Raphanus sativus L.) seeds also point to the temperature-moisture relationship in seed viability studies (Waggoner, 1917; Barrons & McLean, 1945).

Although we do not usually think of true seeds of potato in connection with the propagation of this vegetable, it is important to be able to store such seeds to maintain a certain stock, or to protect hybrid seeds. Cold storage of potato seeds at 32°F. (0°C.) was decidedly superior to storage at room temperature, and sealing was beneficial (C. F. Clark, 1940). At low-temperature storage, the germination of potato seeds remained unimpaired for thirteen years and they still gave 17 per cent germination after twenty years (Wollenweber, 1942). High relative humidity or sudden changes in atmospheric conditions are thought to be detrimental to the viability of potato seeds—hence the value of sealing (Stevenson & Edmundson, 1950). Similar results from storage have been obtained by Haigh (1952) and C. H. Coleman (1953), both of whom report that some potato seeds show a dormancy when freshly harvested and may not reach full germination capacity until 4 to 8 weeks later. Okra seeds with 8 per cent moisture content can be stored successfully for at least seven years at 35° to 40°F. (J. A. Martin & Crawford, 1954).

Deleterious effects of high moisture and temperature have been demonstrated for carrot seeds stored for as little as ten days at 30° to 40°C. and with 13 per cent of moisture (Aki & Watanabe, 1955).

Thus we have seen that low moisture content and low temperatures —especially temperatures below freezing—are best for extending the life-span of vegetable seeds. The harmful effect of below-freezing temperatures for some seeds, however, has been pointed out by Weibull (1955), and was discussed in Chapter IV.

Flower Seeds

The rapid deterioration of many flower seeds under ordinary methods of storage has led a number of seedsmen and florists to publish notes at various times on their own personal experiences. These notes have not usually included experimental data, but, for the most part, have simply emphasized the need for well-planned experiments from which reliable storage methods could be determined. Bodger Seeds, Ltd., El Monte, California, U.S.A., issued a bulletin in 1935 entitled *Valuable Information for Seedsmen* (Anon., 1935). The bulletin lists many flower seeds in tabular form with information on longevity and germination. No special storage conditions are described, however. Some tests conducted by the California Federal State Seed Laboratory, in which sixty-four samples of flower seeds involving thirty genera were stored for ten years, represent extensive work to obtain necessary data on the longevity of these forms. The results obtained from this experiment have been reported by Goss (1937). Although this report contains much information of vital importance, it does not shed any light on the possible extension of life under special storage conditions, as all of the seeds were merely placed in paper envelopes and stored in a tin box in the seed laboratory at Sacramento.

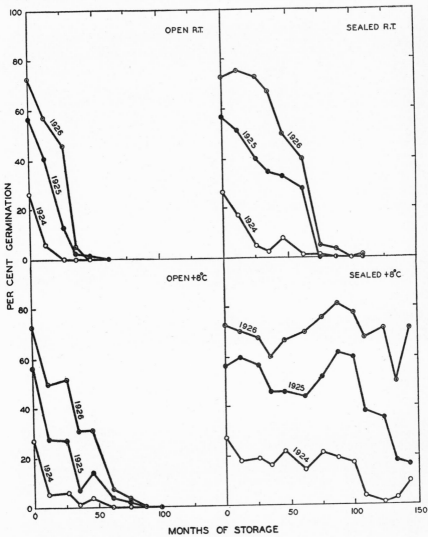

FIG. 8.—Graphs indicating viability of annual delphinium seeds of the 1924, 1925, and 1926 crops, as shown by germination tests on moist filter-paper at 15°C. after storage in open or sealed containers at room temperature (R.T.) or +8°C. for different lengths of time after December 1926.

Delphinium seeds, especially those of the perennial forms, lose their viability rapidly under ordinary conditions of storage. Increased interest in delphinium culture has served to emphasize this problem. Annual and perennial delphinium seeds of the 1924, 1925, and 1926 crops, were placed under special storage conditions in December 1926 (Barton, 1932, 1933,

1939*b*, 1953*a*). This means that two-year-old, one-year-old, and fresh seeds were stored. The storage temperatures used were room temperature, 8°C., and −15°C. up to 1933, after which the 8°C. storage lots were placed at 5°C. and the below-freezing storage was at −5°C. Sealing of these seeds was effected in small glass vials with cork stoppers covered with paraffin. The sealed vials were then placed in large bottles fitted with ground-glass stoppers.

Fig. 8 shows the effect of open and sealed storage at room temperature and 8°C. on the keeping quality of annual delphinium seeds stored for 0 to 143 months. In their responses the three seed-crops showed similar trends throughout, despite the fact that the 1924 crop seeds gave an initial germination of only 27 per cent as compared with 57 and 72 per cent for the 1925 and 1926 crops, respectively. Open storage at room temperature proved most deleterious, causing rapid falling off in viability until all the seeds were worthless after thirty-five months of storage. When sealed containers were used, however, the decline in the power to germinate was less rapid. Germination after sixty-two months in sealed containers at room temperature compared favourably with that obtained after twenty-six months in open containers. Seeds in open storage at 8°C. fared slightly better than those in open storage at room temperature but did not 'keep' as well as those sealed at room temperature—especially after eleven months of storage. Sealed storage at 8°C., however, served to keep these seeds viable for 143 months.

In Fig. 9 is depicted the behaviour of perennial seeds under similar conditions. Here germination of seeds from −15°C. storage has been shown rather than of those from 8°C., as the superiority of the lower temperature was demonstrated early in the storage period. Also, only the 1925 and 1926 crops are pictured as the 1924 crop was worthless when received in December 1926. Again, the similar trend in behaviour in spite of initial germination differences is to be noted. Open storage at −15°C. was far superior to open or sealed storage at room temperature for preserving the germination capacity of these seeds. Sealed storage at −15°C. proved best of all. Here seeds which were relatively fresh when stored in December 1926, kept their viability unimpaired for 143 months.

Tests made after different periods of storage have shown that these old delphinium seeds were not only capable of germinating on filter paper but also produced seedlings in soil (Barton, 1939*b*). Some of the seedlings thus produced from seeds taken from the best conditions after 123 months of storage were grown to maturity. The appearances of typical annual and perennial plants are shown in Plate 4(*a*). These seedlings compared favourably with those obtained from fresh seed planted at the same time. Seeds produced on the plants from old seeds were viable and when planted produced normal seedlings. The delphinium seeds, both annual and perennial, which were stored in 1926, survived for nineteen years when kept at temperatures below freezing (Barton, 1953*a*).

Deeter (1938) obtained 90 per cent germination from delphinium seeds stored in an envelope at room temperature at Danville, Pennsylvania, U.S.A., for five-and-a-half years. The report of Goss (1937) included both perennial and annual delphiniums. He found that the former deteriorated more rapidly than the latter, though there was some difference in the individual crops. There is no doubt that the life-span of these seeds can be increased by storage under favourable conditions.

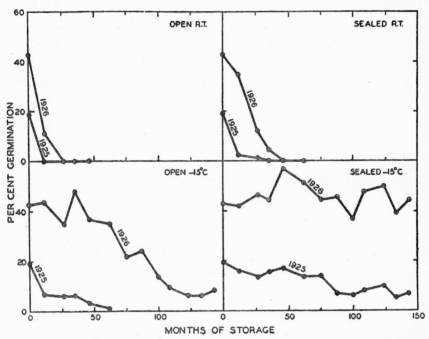

FIG. 9.—Graphs indicating viability of perennial delphinium seeds of the 1925 and 1926 crops as shown by germination tests on moist filter-paper at 15°C. after storage in open or sealed containers at room temperature (R.T.) or −15°C. for different lengths of time after December 1926.

Seeds of an aster (*Callistephus chinensis* Nees) were stored, using three temperatures (room, 5°C., and approximately −4°C.), three moisture contents (from about 4 to 8 per cent), and open and sealed storage (Barton, 1939*b*). Two types of storage containers were employed: hermetically sealed glass tubes which were not opened from the beginning of the experiment until the viability tests were made, and tin cans sealed with sealing wax which were opened at the time of each test and re-sealed. To maintain viability of these seeds in open containers for periods of longer than one-and-a-half years, it was necessary, under the general conditions of this experiment, to use a temperature below freezing-point. There was some indication that deterioration was more rapid in open containers in a 5°C. room with a saturated atmosphere than at ordinary room temperature when the humidity was not so high. However, the rate of decline in germination power in the two latter storage conditions was similar, the seeds being worthless in both instances after storage periods of longer than ,one-and-a-half years. When dry seeds were kept in sealed containers, the length of life at both room temperature and 5°C. could be prolonged—at the latter temperature for as long as nine years. A temperature of −4°C. permitted retention of viability for at least sixteen years in open or sealed storage (Barton, 1953a). Verbena seeds have a shorter life-span than aster under the above conditions.

Winter-flowering sweet pea (*Lathyrus*), pansy, Giant Golden Queen (*Viola*), and *Venidium* seeds, were stored both air-dry and with reduced moisture contents at room temperature, 5°C., and approximately −4°C. (Barton, 1939b). The limited supply prevented extensive tests being made with seeds of the sweet pea, but they remained viable under all conditions for thirteen months, and in sealed containers for thirty-three months. Limited tests with pansy seeds indicated that open storage in a saturated atmosphere at 5°C. was harmful, the seeds being much reduced in viability after nineteen months of storage. Open storage at room temperature permitted retention of viability, though this was somewhat reduced, up to thirteen months, when the seed supply was exhausted. Sealed storage at room temperature kept the seeds viable for nineteen months but proved deleterious when the period was extended to thirty-three months. Under the conditions of Barton's experiment, pansy seeds kept best at 5°C. in sealed containers. Ball (1935, p. 49), writing of pansy seed, stated that 'under favorable conditions its viability will extend over two years, but no responsible seedsman depends on this'. He further remarked that storage conditions must be favourable in order to keep the seeds in good shape until the August after harvesting, but he did not describe these favourable conditions. Harrold (1935) believed that some of the failures attributed to poor pansy seeds were really caused by high soil temperatures or a hot dry atmosphere, pointing again to the importance of adequate testing procedures. Goss (1937), reporting on the results of storing pansy seeds in paper envelopes, presented data which showed 45 per cent germination after five years of storage and 2·5 per cent after nine years. The results of other tests (Barton, 1939b) also indicated that pansy seeds may be kept viable for periods of longer than two years, and probably their life-span may be increased several-fold by sealing seeds with approximately 5 per cent moisture for storage at a low temperature.

Venidium seeds of poor quality maintained their vitality for thirty-seven months under a variety of storage conditions (Barton, 1939b).

Over a ten-year period, Heit (1957) tested the germination responses of seeds of forty-two types of flowers in the laboratory, greenhouse, or field, and concluded that germination failures are not always caused by weak or old seeds, as adverse weather and soil conditions, poor cultural practices, insects, and diseases, are all important factors in survival.

It has been generally recommended that lily seeds be planted while they are fresh. Craig (1931), however, stated that *Lilium auratum* will start better from year-old seeds than from fresh seeds. He also found that *Lilium regale* seeds which had been stored for five years at Weymouth, Massachusetts, germinated nearly 100 per cent. Clement (1938) reported more rapid loss of viability of *L. regale* seeds. He obtained about 80 per cent germination from 1936 and 1937 crops planted on 12 April 1938, but he secured only 8 per cent germination from the 1935 crop treated in the same way. Thus he found that these seeds kept for two years but were practically worthless after three years. Neither Clement nor Craig described the storage conditions employed. It is presumed, however, that open storage under ordinary room conditions of temperature and humidity was used.

Seeds of the regal lily (*Lilium regale* Wilson) have been kept under various storage conditions at the Boyce Thompson Institute for fifteen years, germination tests being made at least once a year (Barton, 1939*c*, 1948). After completion of preliminary germination tests, moisture determinations were made preparatory to storage. Air-dry seeds were found to contain 9·9 per cent moisture, calculated on the basis of the dry-weight of the seeds. One lot of seeds was placed in a desiccator over calcium oxide for two days, after which they were stored. Determinations made on this second lot showed 4·5 per cent moisture. Seeds with this reduced moisture content as well as air-dry seeds were stored in sealed containers in the laboratory, at 5°C., and at approximately −5°C. Air-dry seeds were also stored in open containers. Sealing was effected by using tin cans with tight-fitting lids sealed with sealing wax. Tests for determining viability were made in moist granulated peat-moss at a daily alternating temperature of 15° to 30°C., and in soil in the greenhouse. In the case of the alternating temperature, the cultures were kept at 15°C. for sixteen hours and at 30°C. for eight hours each day. Germination at the controlled alternating temperature was practically complete after two weeks, while a slightly longer period—up to twenty days—was required for the completion of seedling production in the greenhouse.

As there was very good agreement between germination at controlled temperature and seedling production in the greenhouse, data for the latter only are presented. Also, results of viability tests after 1, 2, 3, 5, 7, 11, 13, and 15 years were selected as representative of deterioration under the various storage conditions. These data are shown in Table IX.

It will be noted that air-dry seeds remained fully viable in the laboratory for one year but were reduced in germination capacity after two years, none being viable after three years. Sealing had no appreciable effect on the seeds under these conditions. However, when part of the moisture was removed and the seeds were placed in sealed containers, good germination could still be obtained after five years of storage in the laboratory.

At 5°C., open storage proved unsatisfactory for keeping seeds viable. No tests were made up to the three-year period but at this time the germination capacity was so reduced that only 16 per cent produced seedlings. Sealing was effective in keeping seeds with 9·9 per cent moisture viable at 5°C. for eight years, but thereafter there was a gradual decline in germination ability to 8 per cent at the end of fifteen years. With reduced moisture content the longevity at 5°C. was extended.

TABLE IX

REGAL LILY. SEEDLING PRODUCTION IN SOIL IN THE GREENHOUSE AFTER STORAGE. FRESH SEEDS GAVE 86 PER CENT SEEDLING PRODUCTION

Storage conditions			Per cent seedling production after storage for No. of years indicated							
Temperature	Moisture content per cent*	Open or sealed	1	2	3	5	7	11	13	15
Laboratory	9·9	O	88	42	0	0	—	—	—	—
		S	87	35	0	0	—	—	—	—
	4·5	S	91	89	87	74	31	0	—	—
5°C.	9·9	O	—	—	16	0	—	—	—	—
		S	—	—	—	87	87	37	17	8
	4·5	S	89	86	96	94	93	76	49	—
Approx. −5°C.	9·9	O	98	78	95	87	92	80	83	74
		S	91	87	92	90	96	91	85	81
	4·5	S	93	89	96	86	88	91	89	93

* Calculated on the basis of the dry-weight of the seeds.
— Indicates that no test was made.

The effectiveness of 5°C. as a storage temperature is thus demonstrated. The cause of the rapid loss of viability of seeds stored open at this temperature is to be found in the high humidity prevailing in this room. The combination of this lower temperature and high atmospheric humidity proved only slightly superior to the higher temperature but lower humidity of the laboratory for keeping regal lily seeds. When the absorption of excess moisture was prevented by the use of air-tight containers, seeds remained viable for at least eight years at 5°C. When this lower temperature was used for storage, sealed air-dry seeds (with 9·9 per cent moisture) kept well but, at the higher temperature of the laboratory, it was necessary to reduce the moisture content in order to permit retention of vitality for longer than two years (Table IX).

At −5°C. both open and sealed storage kept the seeds viable for fifteen years. There may have been an indication of reduced germination-capacity of the seeds from open storage at the fifteen-year test (74 per cent seedling production), but this remains for future tests to confirm.

Within the limits of this experiment, the moisture content of the seeds stored at a temperature below freezing-point had little or no effect on their keeping quality. As storage periods lengthen, the superiority of below-freezing temperatures over 5°C. for maintaining the viability of many kinds of seeds is being demonstrated. It should be pointed out, however, that approximately −5°C. is the lowest temperature used in these tests. Other experiments will be necessary to determine the effects of still lower temperatures.

Plate 4(b)A shows the storage temperature effects as related to open and sealed containers, while Plate 4(b)B shows the effect of reduction of moisture content on sealed storage in the laboratory, at 5°C. and −5°C. In sealed storage at −5°C., these seeds of regal lily remained fully viable for fifteen years (Barton, 1953a).

The results reported here are in agreement with those obtained by Clement (1938), who probably had storage conditions in Ontario similar to the open laboratory at Yonkers. A drier atmosphere could have accounted for the extension of the life of regal lily seeds to five years as reported by Craig (1931).

In conclusion it may be said that if a dry cold-storage room is available, seeds of the regal lily can be kept fully viable for at least ten years. If the atmospheric humidity of the room is high, it is necessary to seal the seeds (which should be thoroughly air-dry, i.e. contain less than 10 per cent of moisture) unless a temperature below freezing-point can be used. If no cold-storage chamber is available, the seeds may be kept at room temperature for five years by reducing their moisture content to approximately 5 per cent and placing them in sealed containers.

Germination tests of air-dry *Gladiolus* seeds stored under controlled conditions have shown that they can be stored successfully for twelve years at −4°C., or at 5°C. if the container is not opened in the interim (Barton, 1953a).

The viability of *Gentiana crinita* Froel. seeds was determined after storage under various conditions (Giersbach, 1937). Seeds stored in open containers at room temperature and 5°C. showed a complete loss of viability after one year. Seeds sealed in air at room temperature had also lost their germination capacity by this time, whereas those sealed in a partial vacuum at room temperature and those sealed in air or a partial vacuum at 5°C. still retained their original germination power.

Knudson (1941) was able to store seeds of *Cattleya* for fourteen years and still obtain germination, but seeds of other orchids could not be kept so long. The same general conclusions concerning orchid seeds were reached by Svihla & Osterman (1943), but they found further that these minute seeds could be frozen and dehydrated and still remain viable. Seeds of the tuberous begonia and snapdragon, which are two to ten times the size of orchid seeds, were similarly treated but failed to survive.

In our enthusiasm for storage at low temperatures we should not lose sight of the possibility of damage to some seeds. Weibull (1955) conducted experiments on the storage of flower seeds from 1952 to 1955 and found that they showed very definite reactions to storage at −20°C. He classified the seeds studied into four groups, as follows:

1. −20°C. was favourable for storage of *Kochia scoparia*, *Myosotis sylvatica*, and *Callistephus chinensis*.
2. −20°C. seemed favourable, but the results were less clear, for *Dahlia variabilis*, *Gypsophila paniculata*, *Primula malacoides*, and *Zinnia elegans*.
3. No difference between −20°C. and 'ordinary storage' was found for *Chrysanthemum spectabile*, *Delphinium hybridum*, *D. ajacis*, *Verbena venosa*, and *Salvia splendens*.
4. −20°C. seemed to be disastrous for *Antirrhinum majus*, *Asparagus plumosus*, *Petunia hybrida*, and *Viola tricolor* var. *maxima*.

It is possible that this experiment was not continued long enough for the superiority of the below-freezing temperature to become evident in categories 2 and 3, as it often is not manifested until the seeds have been in storage for five years or more (Barton, 1953*a*).

VII

LONGEVITY OF SEEDS OF FIELD CROPS

WHEAT

THE keeping quality of wheat grain is of great interest, not only because good seeds are vital to good crops, but also because of the food value of the grain. Conditions which keep the seed good for planting purposes also maintain good food value and taste.

Carruthers (1911) was one of the first to record the germination of wheat seeds of different ages. He tested so-called white and red wheat seeds of the 1896 crop each year for nine years. Fresh seeds gave 100 and 99 per cent germination, respectively. After five years of storage these percentages decreased to 88 and 80. Extension of the storage period to eight years caused a further reduction in viability, so that only 29 and 51 per cent of white and red wheat seeds, respectively, were still alive at the end of this time. J. White (1909) recorded 2 per cent germination of a South Australian wheat seed-lot which was sixteen-and-a-half years old.

Although these early records indicate that wheat has a short-lived seed, later controlled experiments have shown that the life-span can be extended materially by storage under controlled conditions. One of the first of these to be instigated was that of Whymper & Bradley (1934), who stored fresh seeds of Rivet wheat at room temperature in sealed tubes under a variety of conditions. No germination tests were made on these seeds until 1927, fifteen-and-a-half years after the commencement of the experiment, when only one lot was found to be viable. This lot was of seeds placed in tubes with dry calcium chloride.

Further storage was made of seeds of two wheat varieties after drying in a desiccator over calcium chloride (Whymper & Bradley, 1934). Desiccation of the variety 'Square Heads Master' to 4·10 per cent moisture, followed by sealing and storage in the laboratory in the dark, preserved viability for up to about nineteen years. Highly desiccated Rivet wheat seed (moisture content of 0·66 per cent) still showed germination of 82 per cent at the end of fifteen years and nine months of storage. In some of the experiments air was excluded by covering the seeds with wax, immersion in oil, or sealing in an atmosphere free from oxygen. The authors state that it appears from the results that these methods caused cessation of respiratory exchange, resulting in death of the seeds, whereas restriction of respiration by desiccation extended viability by the retardation of intracellular reactions. Toxic substances may have accumulated in the seeds when gaseous exchange was restricted. The authors predicted that the life of wheat seeds may, under the best conditions they used, be approximately fifty years. A later release on the performance of these same seeds after twenty to thirty-two years of storage showed a decline in germination from 82 per cent after twenty years to 69 per cent after thirty-two years (Whymper & Bradley, 1947). Under the experimental conditions used, wheat seed with a moisture content of 4·3 to 4·8 per cent dropped in germinative power 17 per cent in nineteen years and 31 per cent in thirty-one years, or about 1 per cent per year. At this rate, the authors point out, their estimate of fifty years for the life-span of the seeds is likely to be too low. Unfortunately, the seed supply has been exhausted and so no more tests can be made.

D. W. Robertson and co-workers in a series of publications (D. W. Robertson & Lute, 1933, 1937; D. W. Robertson *et al.*, 1939, 1943) emphasized the importance of humidity on wheat seed longevity. They found that length of life increased as the humidity was decreased. In an unheated room in Colorado, germination decreased 7 per cent in the first ten years of storage, and by the end of twenty-one years only 12·8 per cent germination was obtained. These seeds contained 9·5 to 11·4 per cent of moisture as compared with only 4·3 to 4·8 per cent in the seeds stored by Whymper & Bradley. The higher the moisture content, the shorter was the life-span and the greater the injury from exposure to high temperatures (Whymper & Bradley, 1934*a*, 1934*b*).

Swanson (1941) has also demonstrated the relation between temperature and moisture content, and recommends that high-moisture wheat grain be stored at a low temperature. Wheat seeds were placed in 4-oz., stoppered bottles (1) in the laboratory and (2) at 41 °F. Germination after laboratory storage of seeds containing 10 per cent of moisture was 92 per cent after eighteen weeks of storage. Increasing the moisture content to 14 per cent decreased the germination to 51 per cent after eighteen weeks of storage, and a further increase to 16 per cent of moisture resulted in the death of all the seeds after only nine weeks. On the other hand, the germination capacity remained unaltered upon storage for up to eighteen weeks at 41 °F.—regardless of the moisture content, which was from 10 to 20 per cent.

In 1938 wheat grain was stored in Japan at −25° to −35 °C., for one year only (Tejima & Misonoo, 1938). Viability was good at the end of that time but no further report is available. In the light of other favourable responses to below-freezing temperatures, these wheat seeds should have a long life-span.

In addition to loss of viability, high moisture content of wheat seed causes heating when the grain is stored in large lots; it also increases mould growth and brings about deterioration of baking qualities of the flour (C. H. Bailey, 1917, 1917*a*; McCalla *et al.*, 1939; Lafferty, 1942; Anon.,

(*a*) Delphinium plants produced from seeds stored for 123 months before planting. A, Annual plants; B, perennial plants.

PLATE 4

(*b*) Development from regal lily seeds sown in flats after four and a half years of storage. A, Flats sown with air-dry seeds stored in the laboratory (*left*), at 25°C. (*centre*), and at −5°C. (*right*): (1) in open containers, and (2) in sealed containers. B, flats sown with seeds dried over calcium oxide before storage at various temperatures: (1) in the laboratory, (2) at 5°C., and (3) at −5°C.

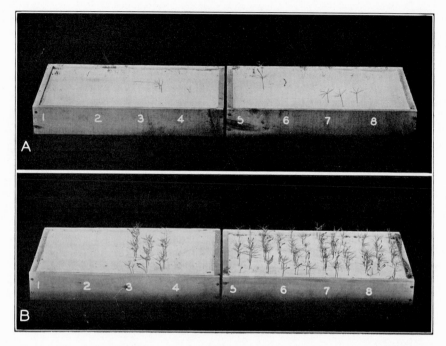

PLATE 5

Pinus taeda: seedling production in the greenhouse after storage for five years (air-dry seeds). A, Planted in flat directly from storage flasks: (1) open at room temperature in dark, (2) sealed at room temperature in dark, (3) in vacuum at room temperature in dark, (4) in vacuum at room temperature in light, (5) open at 5°C., (6) sealed at 5°C., (7) in vacuum at 5°C., (8) open in refrigeration room (−5° to −15°C.). B, Planted in flat after one month in moist granulated peat-moss at 5°C. Numbers as in A.

1945; E. P. Carter & Young, 1945; Milner *et al.*, 1947, 1947*a*; Harris & Walster, 1953; Oxley & Hyde, 1955). Storage conditions maintaining full viability also retain desirable food characteristics.

MAIZE

Developments in the growing, harvesting, and marketing of maize (there known as 'corn') in the United States necessitates a carry-over of large surpluses of grain from one year to the next as a means of price and supply stabilization. Also, the special techniques, time, and labour involved in the production of the homozygous inbred lines necessary for subsequent production of superior hybrid maize, make it imperative that good storage conditions for maintaining viability of breeding-seed stocks as well as planting material be known. Most of the literature on maize is of American origin, because America is the greatest 'corn' producer. Argentina is also important in world production, but the maize grown there is of the 'flint' type and, because of the dry autumn climate, dries down to moisture limits that are safe for storage and shipment. In other parts of the world where maize seed is grown, it is either dried carefully by the individual grower who has a small supply, or it is consumed without the necessity of storing.

In a book by Huelsen (1954), a general account of the taxonomy, morphology, and physiology of sweet corn will be found. He devotes a chapter to the factors, including storage, affecting germination and growth of the seed.

Detailed experiments with Narrow Grain Evergreen sweet corn stored under different conditions of temperature and humidity have been reported by Boswell *et al.* (1940). Moisture and germination tests were made at intervals of ten days for up to 251 days of storage. The original germination capacity of the seeds was 82 per cent. Estimated germination values are given, based on the trend of the curve for successive tests. Storage at 80°F. in humidities of 78, 66, and 44 per cent resulted in 15, 68, and 70 per cent germination, respectively, after storage for 110 days. Seeds stored for a similar period at 50°F. and in 81, 66, and 51 per cent humidity germinated 65, 70, and 72 per cent, respectively. No germination was obtained after storage for the full 251 days at the high humidity and temperature. At low humidity and temperature significant deterioration was evident, but the seeds still gave 72 per cent germination after 251 days. The short total time of storage prevented the appearance of deterioration under all storage conditions, but favourable moisture and temperature for safe storage for short periods are indicated.

Yellow Dent maize seeds of the 1921, 1922, 1923, and 1924 crops were stored in an unheated room in Colorado (D. W. Robertson *et al.*, 1943). Samples were taken for germination in February of each succeeding

E

year. Viability gradually declined from the first to the twenty-first year, the seeds germinating 32 per cent at the end of the period.

A more detailed test was started by Sayre (1947) in 1933. Seed maize with the moisture content adjusted to 5, 7·5, 11, 14·6, or 18·2 per cent was placed in large glass culture tubes. The tubes were then stored at 86°, 72°, 37°-39°, 16°-28°, or 0°F. In 1938, another series was started, using tin cans for storage. Germination tests were made each year for each series. All of the seeds with the highest moisture content of 18·2 per cent which were stored at room temperature lost the power to germinate before the end of one year. Seeds were more tolerant to the higher moisture contents as the storage temperature was lowered. The same author had shown previously (Sayre, 1940) that low temperatures (3° and −25°C.) and absence of oxygen from the storage container extended the life of maize seeds.

Confirmation of the roles of humidity and temperature in determining the keeping quality of maize seed is available in the works of Duvel (1909), Brunson (1946), Swaine (1957), and many others. Drying seed maize with electricity (Duffee, 1937) or with different drying agents (Larter, 1947) before storage has been found effective.

The yield of high quality open-pollinated maize declined gradually after the second year of storage, but still retained satisfactory viability and vigour after five years (Dungan, 1940). Limited tests made by Dungan indicated that some hybrid seed maize produces a better yield than others after one year of storage.

Haber (1950) also demonstrated that there is a genetic basis for longevity of seeds of sweet corn, while confirming that humidity and temperature are the major environmental factors responsible for maintenance of germination capacity. His experimental results showed that, in general, the longevity of hybrid seeds was better than that of the two inbreds used to produce the single-cross hybrid. When two inbreds with good longevity of seeds were crossed, the hybrids had the same character. This was also true of inbreds with short life-spans. Long-lived inbred×short-lived inbred hybrid seeds retained viability much longer than the short-lived parent. It would be necessary to store inbreds with a short-lived seed character under favourable, controlled storage conditions in order to maintain such stock.

Enough work has been done on seeds of maize and of other plants to prove the value of below-freezing storage for long life (see Chapter IV). Advantage is being taken of this knowledge to prepare cold-storage rooms at about 0°F. for the maintenance of valuable maize seed stocks (Anon., 1949, 1950). McRostie (1939) made a study of the thermal death-point of maize from low temperatures (down to −10°F.). Seeds with moisture contents above 15 per cent suffered severe damage which was greater under fluctuating than under constant temperatures. Safe storage of maize grain requires a moisture content well below 15 per cent.

While the food value of maize grain may be directly related to its viability, the popping quality of popcorn appears to be entirely independent of germination capacity (Stewart, 1936). Ability to pop is retained at least fourteen years—six years beyond the ability of the particular lot of popcorn to germinate.

Other Cereals

The life-span of cereal grains has been reported by various authors as follows: wheat, 6 to 32 years; oats, 6 to 29 years; barley, 2 to 10 years; rye, 9 to 10 years (Burgerstein, 1895; Todaro, 1898; Carruthers, 1911; Sifton, 1920; Percival, 1921; Welton, 1921; D. W. Robertson & Lute, 1933; Walster, 1943; Haferkamp, 1949). An atmosphere devoid of oxygen, or, alternatively, drying and sealing, extends the life of cereal grains (Bussard, 1935). Drying is imperative if losses from fungal invasion are to be avoided in combine-harvested barley grain (MacLeod, 1952). The malting industry has to sell, for feeding purposes, many hundreds of bushels of barley grain annually because of the loss in storage of germinability. The greatest damage to malting barley grain occurs during the early storage period before the grain dries out. Storage conditions in the State of Washington are exceptionally favourable for retention of viability by cereal grains (Haferkamp *et al.*, 1953). Moisture content and respiration are important in the heating and deterioration of oat seed (Bakke & Noecker, 1933).

General discussions of the physiology and biochemistry of grain in storage are given by D. W. Robertson *et al.* (1939), Slusanschi (1939), Kretovich (1945), and Blum (1954). Causes of deterioration in, and practices for storage of, cereal grain are discussed by Oxley (1950a, 1955). Anderson & Alcock (1954), in the detailed book they edited on the storage of cereal grains and their products, state in the preface that millions of tons of cereals are wasted every year through spoilage of various sorts in spite of recent advances in technology.

Grasses

Chewing's fescue (*Festuca rubra* L. var. *commutata* Gaud.) seeds have claimed the attention of several workers because of their economic importance and their short life-span under certain storage conditions. Foy (1934) stated that deterioration of these seeds in transit was due to excessive heat and humidity in the hold of the ship. Viability could be maintained by lowering the moisture to about 5 per cent, or by shipping in cool storage. Seeds stored over calcium chloride in the laboratory for seven years still gave over 90 per cent germination, while undried seeds (about 13 per cent moisture) were dead in two-and-a-half years. Commercial drying could best be done by heat. Twenty-one samples of Chewing's fescue seed stored in the laboratory for from twenty-one to thirty-four months lost 6 to 69 per cent of their original viability (Patrick, 1936). Kearns & Toole (1939a) found that the advantages of shipping these seeds under cool conditions from New Zealand were lost within a few months after arrival in the United States, unless cool storage was continued or the seeds were dried out. Seeds dried before shipment and kept dry after arrival in the United States gave germination percentages after eighteen to twenty-six months' storage at room temperature which were equal to those of non-dried seed on arrival. Tests were made using domestic seed and storing it in sealed containers at $-10°$, $2°$, $10°$, $20°$, and $30°$C., after adjustment of the moisture content to 8 to 14 per cent. Here again the interrelationship of temperature and moisture was shown. The higher the temperature of storage, the lower was the moisture content required for maintaining viability. When the seeds are stored at $30°$C., the moisture content should not exceed 8 per cent; at $20°$C. it should not exceed 10 per cent; at $10°$C. it should not exceed 12 per cent.

Fluctuation in moisture content is injurious to Chewing's fescue seeds in storage (LaPine & Milberg, 1948). Gane (1948a) found that the use of artificial atmospheres of nitrogen or nitrogen+1 per cent oxygen had no noticeable effect on the rate of deterioration, while lowering the storage temperature and reducing the water content of the seeds increased their retention of viability during a four-year storage period. Kearns & Toole (1939) emphasized the need for low temperature in the germination testing of Chewing's fescue seed, especially when the seed is fresh.

As early as 1894, Samek published a table showing the results of germination tests of twenty different forage plants for a period of eleven years. The fourteen species of grasses, etc., included in the study showed decreased germination after five years, the actual amount varying with the species, and all of them produced only occasional seedlings after ten years of storage. The grasses, etc., reported on were: French oat grass (*Avena elatior*), Italian ryegrass (*Lolium italicum*), English ryegrass (*Lolium perenne*), tall fescue (*Festuca pratensis*), sweet vernal-grass (*Anthoxanthum odoratum*), meadow foxtail (*Alopecurus pratensis*), timothy (*Phleum pratense*), orchard grass (*Dactylis glomerata*), blue-grass (*Poa pratensis*), crested dogstail (*Cynosurus cristatus*), fiorin (*Agrostis stolonifera*), sheep's fescue (*Festuca ovina*), hair-grass *Deschampsia* (*Aira*) *flexuosa*, and spurry (*Spergula arvensis*).

Using hulled and unhulled seeds of timothy (*Phleum pratense*) stored in six different warehouses in various districts of Sweden, Esbo (1954) found that unhulled seeds maintained good germination capacity until at least the beginning of the second summer. Hulled seeds declined more rapidly, especially as a result of the temperature fluctuations of the first summer.

Excess moisture should be removed from grass seeds as soon as possible after harvesting, according to Williams (1938).

68

It has been observed that certain native grasses in Australia can disappear for a while owing to drought or animal injury or grazing, and then reappear (Myers, 1940). Seeds of some of these grasses were kept in packets and stored in the laboratory at Sydney under ordinary atmospheric conditions for twelve years. *Panicum antidotale* gave 3 per cent germination after such storage for twelve years, but 88 per cent after seven years; *P. decompositum* gave 3 per cent germination after nine years, but 63 per cent after four years; *P. prolutum* gave 2 per cent germination after eight years, but 65 per cent after four years; *Paspalidium jubiflorum* gave only 1 per cent germination after ten years, but 72 per cent after five years; *Danthonium semiannularis* gave 5 per cent germination after four years, but 100 per cent after one year. The greatest germination in every case was from seeds one or more years old, thus indicating a dormancy in freshly harvested seeds.

Seeds of *Andropogon furcatus*, *A. scoparius*, *Sorghastrum nutans*, and *Bouteloua curtipendula*, are also dormant when fresh and do not begin normal germination until from fourteen to eighteen months after harvest (Coukos, 1944). After dormancy has been broken by storage under dry conditions, germination remains at a high level for several months. Cold storage in sealed jars extends the life-span to at least thirty-eight months—except in the case of *Sorghastrum nutans* seeds, which remained viable for only twenty-four months.

Italian ryegrass seed could be stored successfully for two years if the seeds were dried to 13 per cent of moisture (Gorman & Greenwood, 1951). Longer storage periods required further reduction in moisture content.

One crop of carpet-grass (*Axonopus affinis*) seed remained fully viable for three years of storage while the germination of another crop fell to 5 per cent in three years. The difference may have been due to treatment of the seeds before they were received for special storage (E. H. Toole & Toole, 1939).

For general information on the different phases of grassland cultivation, the reader is referred to a recent book entitled *Grassland Seeds* (Wheeler & Hill, 1957).

LEGUMES

Because of their importance as forage crops, alfalfa or lucerne (*Medicago sativa* L.), various clovers (*Trifolium* spp.), and sweet-clovers (*Melilotus* spp.), have been the subjects of storage investigations. Teräsvuori (1930) obtained some germination of seeds of different species of *Trifolium* which had been stored for fifteen to thirty-four years. Stevens (1935) found that the viability of seeds of alfalfa and sweet-clover stored in the laboratory decreased from about 95 per cent to 60 per cent, and of red clover to 10 per cent, in twenty years. Most of the impermeable seeds of alfalfa became permeable during the first year, but a few remained impermeable after twenty years. Clover seeds became permeable more slowly, and 'hard' seeds persisted for the longest time in sweet-clover. Later reports by Stevens (1943, 1954) included germination data from six samples of alfalfa and two samples each of sweet-clover and red clover seeds stored in ordinary seed envelopes in ventilated metal boxes in the laboratory. Alfalfa seed germinated 37 to 70 per cent after thirty years and showed another 10 per cent decrease at a total age of forty years. All samples of clover seed deteriorated more rapidly, giving only 1 to 13 per cent germination after thirty years. Myers (1952) stated that if lucerne seeds are to be kept in a dry climate, they can be in either open or sealed containers, but to remain viable in a moist climate they must be sealed.

The persistence of viability of seeds of alsike clover after fifteen years of storage was attributed to the 'hard' seeds in the sample (Lafferty, 1931). We have already seen (Chapters I and IV) that the presence of an impermeable coat extends the life-span of many seeds. Similarly, the loss of viability in storage may be caused by a softening of the coats. There are often changes in coat permeability and hence germination capacity between harvest and the time for sowing the following year, as has been found for red clover seeds (J. N. Martin, 1945).

Some leguminous seeds have been claimed to show signs of renewed vigour after having begun to deteriorate (Rodrigo, 1935). This may be related to failure to treat the seed-coats for germination or to seed-coat permeability changes during storage. The date of harvesting determines the moisture content and hence the suitability of seeds of velvet-beans (*Stizolobium* spp.) for storage (Justice & Whitehead, 1942).

Seeds of blue lupine (*Lupinus angustifolius* L.) containing 20 per cent of moisture can be stored without damage for long periods only at relative humidities of less than 70 per cent, and probably only at ones of less than 65 per cent (Brewer & Butt, 1950). High quality, naturally cured seeds of less than 10 per cent moisture content can, however, be stored without great damage for eight weeks at a relative humidity as high as 85 per cent. These seeds were injured by drying artificially to 3·6 per cent moisture, so that their germination was reduced from 85 to 38 per cent, but this low viability was maintained for eight weeks even at 100 per cent relative humidity.

Seventeen different crop seeds were tested by Sonavne (1934) over a period of twelve years. In general, leguminous seeds retained viability longest. The storage of farm seeds by Vibar & Rodrigo (1929) demonstrated the advantage of sealed containers in maintaining viability.

Storage of the important legume, soya bean, will be discussed in the next chapter.

VIII

LONGEVITY OF OTHER ECONOMICALLY IMPORTANT SEEDS

Oily Seeds

soya bean (*Glycine max* Merr.)

In spite of the importance of soya bean production and the relatively short life-span of the seeds under ordinary conditions, detailed work on the requirements for successful storage has been done only recently.

Oathout (1928) was one of the first to report that high temperature and lack of ventilation caused rapid loss of viability in soya bean seeds with moisture contents above 14 per cent. Measurement of the damage to soya bean seeds under the ordinary conditions of temperature and humidity obtaining in Trinidad (monthly maximum from 83° to 87°F., monthly minimum from 65° to 72°F.; humidity at the saturation point almost every night) was made by G. E. L. Spencer (1931). Germination tests of samples stored at 55° to 60°F. were also made. After cool storage for ten months, 100 per cent germination was obtained, while the percentage was reduced to 11 after storage for the same period under ordinary conditions.

Soya bean seeds of eight different varieties, which had been treated with an insecticide, were stored in North Carolina for periods of up to four years and five months (Burgess, 1938). At the end of the storage period, germination varied from 74 per cent for the Biloxi variety to 0 per cent for the Haverlandt (or Herman). Seeds of all varieties germinated from 95 to 100 per cent at the beginning of the test, so varietal characteristics are evidently among the factors determining life-span.

Increasing the moisture content from 9·4 to 19·1 per cent caused rapid loss of viability of soya bean seeds stored at room temperature in Minnesota (Ramstad & Geddes, 1942). Seeds kept in storage for two years or longer in Illinois gave poor field-stands (Burlison *et al.*, 1940). Ottawa, Ontario, Canada, apparently affords a better climate for soya bean seed storage; seeds not more than three years old germinated satisfactorily, though not quite as well as fresh seeds (Laughland & Laughland, 1939). Four-year-old seeds showed considerable weakness and anything older was unfit for planting.

A more comprehensive experiment was started in 1934 to determine the effects of temperature of storage and moisture content of soya bean seeds on their life-span (E. H. Toole & Toole, 1946). Seeds of the soya bean varieties Otootan and Mammoth Yellow were adjusted to moisture contents of approximately 18, 13, 9, and 5 per cent, and placed in sealed glass pint fruit-jars at controlled temperatures of −10°, 2°, 10°, 20°, and 30°C. Germination at the time of storage was 93 to 97 per cent. Tests, made at intervals after storage, indicated a similar response of the two varieties. The seeds with the highest moisture content lost viability rapidly at the higher temperatures but maintained good germination for five to six years at −10°C. Air-dry seeds (approximately 13 per cent moisture) fell rapidly in germination capacity at 20° and 30°C.; there was, however, no loss until after three years at 10°C., and the original viability was retained to the end of the experiment at −10°C. Seeds with approximately 9 per cent of moisture showed slight deterioration after a year at 30°C., but rapidly lost germination power after that time. Lowering the temperature to 10°, 2°, or −10°C., maintained full viability for ten years of seeds with 9 per cent of moisture. Drying the seeds to 5 per cent moisture content caused immediate injury as evidenced by cracks across the cotyledons, severance of vascular connections of some cotyledons, and severance of the plumule.

Other evidences of the deleterious effect of excess moisture on soya bean seed viability, together with additional data on longevity, have been presented by Akamine & Ripperton (1938), Ramstad (1942), Germ (1943), Tervet (1943), Robbins & Porter (1946), Milner & Thompson (1954), and a series of articles by Milner & Geddes (1945, 1945a, 1946, 1946a). Some of these authors also deal with respiration, heating, and moulds induced by excess moisture.

COTTON (*Gossypium* spp.)

Regions where cotton is grown are usually humid with high temperatures—conditions known to be detrimental to most seeds. Cotton-seeds are no exception. For example, in Charleston, South Carolina, cotton-seeds held in bags deteriorated rapidly after two years; but those in which the moisture content had been reduced below 8 per cent, and which had been stored in tin containers, showed only a slight decrease in viability after seven years, and some of these seeds were still germinable after ten years (Simpson, 1942). In a series of publications, Simpson has given data on the cotton-seed storage problem, including the results of a controlled storage experiment begun in 1937 (Simpson & Stone, 1935; Simpson, 1935, 1940, 1942, 1946, 1953, 1953a).

In the controlled experiment, cotton-seeds of two upland varieties were adjusted to levels of moisture from 7 to 14 per cent and stored in sealed containers at constant temperatures of 90°, 70°, and 33°F. Corresponding controls were subjected to normal air temperatures at Knoxville, Tennessee. At 90°F., seeds with 14 per cent moisture were all dead in four months, and those with reduced moisture were either dead or nearly so within three years. All seeds stored at 33°F. for three years 'kept' without appreciable deterioration, but those with 14 per cent moisture began to deteriorate after five years at this low temperature. At lower moisture contents, seeds at 33°F. maintained viability after seven years. Seeds stored at 70°F. were intermediate in response between those at 90° and those at 33°F. Seeds at 33° and 70°F. were continued in storage for up to fifteen years, when the driest lot (7 per cent moisture) still gave 73 per cent germination (after being at 70°F.) and 91 per cent germination (after being at 33°F). Seeds with 13 per cent moisture which had been stored at 33°F. were also still viable after fifteen years (72 per cent), but those with 14 per cent moisture content were all dead by then. Simpson states (Simpson, 1953a, p. 391): 'The much longer life under storage at 33°F. plus the striking differences in longevity obtained for moisture levels within and between the other storage temperatures, furnish conclusive proof of the inter-relationship of moisture and temperature in cotton seed deterioration. The moisture and temperature tolerances determined in this experiment are sufficiently definite to serve as a guide in safely storing cotton seed. If the indicated conditions of moisture and temperature are maintained, good seed may be stored for fifteen or more years with reasonable assurance that viability will be retained.'

In work carried out at the Hellenic Cotton Research Institute, Sindos, Greece, Christidis (1954) used seed of the same genetic constitution produced in successive years but all tested during the same year. Three varieties of upland cotton served as source material. The experiment showed that the genetic constitution remained unchanged in the individual cases, although there was variation in their behaviour, owing not only to the age of the seed and the error of sampling but also to the conditions under which the seeds were produced. In spite of these variations, however, seeds of all three varieties germinated satisfactorily after eleven years of storage.

Whether controlled storage conditions must be used for cotton-seed depends, as for many other seeds, upon the locality of storage. Thus it is necessary to dry the seeds and store in sealed containers at the Central Experiment Station, Bureau of Plant Industry, in Manila, Philippines (Flores, 1938). Sun-drying before storage is effective. A consistent decrease in germination capacity was found in cotton-seeds stored under warehouse conditions in California, so that the decline was significant in three years (Towers & Harrison, 1949). An investigation of the causes of deterioration of cotton-seed in the equatorial province of the Sudan (Anthony & Tarr, 1952) showed that loss of germination capacity was closely related to the relative humidity. The optimum humidity for storage was found to be about 33 per cent. This was much better than higher humidities and somewhat better than a dry atmosphere, which indicates that excess drying is injurious. Phillis & Mason (1945) also found that there was a surprising loss in viability of cotton-seeds stored under conditions of very high aridity. However, if seeds were transferred from o to 50 per cent relative humidity for six months before they were germinated, their performance was improved over that produced by planting the seeds directly from the low humidity storage. The failure to germinate after extreme desiccation is apparently not due to the development of impermeable seeds, as such seeds swelled readily when moistened and some of them produced short radicles. However, these seeds died without further development. Artificial drying of cotton-seed at the gins extended the life-span and prevented the development of free fatty acids (Rusca & Gerdes, 1942).

Formation of free fatty acids in stored seeds, and the use of chemicals for desiccation and extension of life of cotton-seed, are discussed in Chapter V. Infection by bacteria or fungi was not associated with loss of viability in tests conducted by Arndt (1946).

FLAX (*Linum* spp.)

Moisture and temperature have again been the factors most studied in connection with this and other oily seeds. In North Dakota, flax-seeds from seven to fourteen years old gave satisfactory germination but samples from fifteen to eighteen years old showed loss of viability, germinating from 50 to 89 per cent (Dillman & Toole, 1937). Germination of fresh seeds depended upon the particular harvest. A wet harvest and threshing season caused damage to the seeds. Also, seeds of high moisture content (from 10 to 18 per cent) were injured. Flax-seed of good quality, stored under favourable conditions, could be used for seeding for a period of six or eight years.

Decker & Reitz (1948) stored Linota flax-seed of high germinating power at 40°F., about 72°F., and 86°F., after adjusting the moisture content to three different levels (9, 11, and 13 per cent). Storage was for a period of 946 days, with periodic sampling for laboratory and field plantings. Conclusions from the data obtained were: that flax-seed cannot be stored safely at 86°F. at moisture levels of 9 per cent or above; that moisture contents above 11 per cent caused rapid deterioration after ninety days in storage at 72°F.; and that a temperature of 40°F. was best, regardless of the moisture content.

High moisture content and injured seed-coats influence the oil quality of flax-seeds (Painter & Nesbitt, 1943; Kopeĭkovskiĭ *et al.*, 1956).

PEANUTS (*Arachis hypogaea* L.)

Well-cured peanut seeds held in the shell under room temperature conditions kept well for at least five years and the quality was not improved by storage at 32° or 40°F. (Beattie, 1931). Shelling the seed before storage caused slightly greater decline in germination capacity. Similar results were reported later by Beattie *et al.* (1932). Excess drying before shelling increases kernel breakage (Beattie & Kushman, 1947). Mathur *et al.* (1956) recommend that shelled peanut seeds should be stored at 32° to 35°F., and unshelled ones at room temperature. G. E. L. Spencer (1931) reported a small but significant difference, in favour of the cool storage, in the germination of seeds held in cool and ordinary storage.

CASTOR OIL (*Ricinus* sp.)

The quality of seeds of *Ricinus* declines rather rapidly during the first few months after harvest and then more slowly, so that 75 per cent of their original vitality is still present after ten years (Guillemet, 1931).

OTHER SEEDS

Whether maintenance of the viability of tobacco (*Nicotiana* sp.) seeds is a problem also depends upon the place of storage. It was reported early in this century (Shamel & Cobey, 1907) that thoroughly dried tobacco seeds could be kept with perfect safety for ten years or even longer. Similar results were obtained in Wisconsin (J. Johnson *et al.*, 1930) and in Connecticut (Anon., 1936). In places like Florida and Puerto Rico, however, seeds more than two years old are apt to produce inferior stands.

Poptzoff (1933) recommended 55 to 60 per cent relative humidity and a temperature not greater than 15°C. for the storage of tobacco seed. According to Poptzoff, excess drying does not injure the seeds, but rather prolongs their life-span. Immature seeds do not keep as well as ripe ones, which should be stored at low relative humidity and a low temperature above freezing-point (Chirkovskiĭ, 1938).

Kincaid (1943) controlled the moisture content of tobacco seeds by storing them over chemicals which gave an atmosphere of about 1, 40, 60, and 100 per cent relative humidity—or, in some cases, seeds were sealed in glass tubes after their moisture content had been adjusted. Test lots were kept in the refrigerator at about 5°C., and in a basement, a laboratory, and an attic. After eleven years, seeds stored over calcium chloride (1 per cent relative humidity) at each location, and all of the seeds in the refrigerator, germinated well. Samples under normal, uncontrolled conditions in the laboratory were nearly all dead after three years. Seeds with 5·3 per cent or less of moisture, which had been stored in stoppered vials, had deteriorated markedly after eight years of storage—indicating a harmful effect of excessive drying which was not found by Poptzoff (1933).

Twenty-year-old seeds of tobacco stored dry, in the absence of air and light, were still fully viable (Schloesing & Leroux, 1943).

Sugar beet seeds are significantly reduced in germinating power after storage for five years (E. W. Schmidt, 1934) as shown by germination percentages and the poor root systems of resulting seedlings. According to Filutowicz & Bejnar (1954) the limit of safe storage of these seeds could be extended to seven years (for eleven varieties harvested for seven successive years) without decreasing their planting value.

Mesquite (*Prosopis velutina* Woot.) seeds may remain viable for forty-four years owing principally to impermeability of the coats (S. C. Martin, 1948).*

Seeds of coltsfoot (*Tussilago farfara* L.), which ordinarily lose their viability within four months at any temperature above 0°C., kept it for more than two years at −15°C. (Dorph-Petersen, 1928a).

Seeds of *Hibiscus cannabis* L. lose their viability quickly—even under the most favourable storage conditions, according to Poptsov (1929).

Moisture content is important in the safe keeping of sorghum seed (Shedd & Walkden, 1947), but these seeds will maintain 50 per cent of their original viability in certain localities without special storage (Karper & Jones, 1936).

Seeds of prickly-pears (*Opuntia* spp.) germinated after forty-four weeks, including twelve weeks of storage and thirty-two weeks under conditions suitable for germination (Myers, 1939). Later tests indicated that *Opuntia* seeds may remain viable for years (Myers, 1939).

Arnica montana L. seeds maintain their viability for a year in dry storage (Esdorn, 1940). Poor quality *Arnica* seeds require three weeks for germination, while good ones germinate in two weeks.

Information is available on a few horticultural seeds not dealt with in other parts of this book. Among these are seeds of *Nandina domestica*, which can be stored dry at a low temperature for ten months without appreciable loss of viability (Afanasiev, 1943). Testing of these seeds is complicated by their delayed germination which results from a rudimentary, undifferentiated embryo.

Limited tests with the seeds of *Angelica archangelica* indicated that sealing in an air-tight container stored at about 35°F. kept the seeds viable for eight months, at which time the experiment was terminated (Taylor, 1948). These seeds also required a special germination temperature of 42°F. for sixteen hours daily and of 77°F. for the remaining eight hours.

Data are available for *Kochia indica* seeds which indicate that they retain their germination capacity after storage for various periods of time when open to the air in the laboratory. Conditions tried include: open at 30°C., open at 38°C., and sealed over calcium chloride at 30°C. (El-Shishiny, 1953). No deterioration took place for more than fourteen months at 30°C. over calcium chloride. The author states that rapid deterioration of seed in open storage at laboratory temperatures and at 30°C. shows that loss of viability is accelerated by moisture more than by temperature. No doubt the life-span could have been further extended by low-temperature storage of the dry seeds.

* The related Arabian 'Shok' (*Prosopis stephaniana*) also has very hard impermeable seed-coats. Dormancy of the seeds is readily broken by filing or treatment with concentrated sulphuric acid, as was recently demonstrated by Dr. A. K. Khudairi of the Department of Botany, University of Baghdad, Iraq.—Ed.

Blueberry seeds were cleaned in a Waring Blendor and stored dry in vials with cork stoppers in an ordinary refrigerator at about 40°F. (Darrow & Scott, 1954). Good germination was secured from one twelve-year-old lot and from two nine-year-old lots stored in this way. No other special storage conditions were used.

There is a widespread belief that strawberry seeds are very short-lived but they still gave 81 per cent germination after two years of storage in screw-top jars under normal conditions for Sydney, Australia, with a mean temperature ranging from 54°F. in July to 71·8°F. in January, and mean relative humidities of 65 to 74 per cent for the same periods (Myers, 1954).

IX

LONGEVITY OF TREE SEEDS

FOREST trees, especially conifers, are sporadic in their seed production (Wakeley, 1931, and others). The quantity of the harvest, and the intervals of time between seed-bearing, both vary with the different species and with the climatic conditions—including the ease or difficulty of pollination at the time of opening of the inflorescence. Such circumstances make it imperative to use storage conditions which will ensure a constant supply of seeds for planting.

CONIFERS

A considerable amount of work has been done on the viability of coniferous seeds. Cieslar (1897) found that sealed storage of *Picea excelsa*, *Pinus austriaca*, and *Pinus sylvestris* lengthened the life of the seed. The difference in germination of seeds from sealed and open storage after six years was found to be as high as 33 per cent. Haack (1909) confirmed some of the results of Cieslar. He found that sealed as against open storage reduced the fall in germination power of pine seeds by 16 to 68 per cent in different samples in three years. He also found that the seeds must be dried thoroughly if they are to be stored at moderately high temperatures. Drying, however, had no effect on sealed storage in a cellar or an unheated room, but open storage was possible only in a dry room. Haack also emphasized the necessity of a distinction between germinating percentage and the percentage of actual plant production. As the ability to germinate on a synthetic substrate rose, the power to produce plants in soil rose even more rapidly. This effect became more marked when the seeds were of poor quality, or when seedling bed conditions were unfavourable. Only quick-growing seeds produced plants which survived in the seed-bed.

It was reported as early as 1880 (Sargent, 1880) that ten-year-old seeds of *Pinus contorta* would germinate provided they had remained in the cone.

Coker (1909) referred to this remarkable retention of viability in still unopened cones of *Pinus attenuata* and some other forms, and the presence of good seeds in twenty- to thirty-year-old cones. He further pointed out that the conditions existing in serotinous cones are almost ideal for the preservation of seed viability because of the regulation of gas exchange and humidity. Additional proof of this was given by Blumer (1910). Tillotson (1921), working on *Pinus ponderosa*, *P. monticola*, *P. strobus*,

P. contorta, Picea engelmanni, and *Pseudotsuga taxifolia,* spread out all the seeds thinly on a floor and fanned them steadily for two days with an electric fan, after which he stored them under various conditions. He also found that storage in air-tight containers was far superior to any other form. In this condition seeds were little affected by temperature. However, contrary to the findings of other workers, he reported the poorest results from low-temperature storage.

In 1928 an anonymous article (Anon., 1928) on the keeping-quality of noble fir seed (*Abies nobilis*) reported the beneficial effect of cold storage. These seeds were well dried before sealing in air-tight glass jars. Wakeley (1931*a*) stored longleaf pine seeds in tight containers at low temperatures and found that they germinated well one or two years after collection. Seeds of '*Sequoia gigantea*' were stored effectively though with decreasing viability for eighteen years in an unsealed Mason jar at room temperature (Toumey, 1930). On the other hand, the same worker reported little or no germination of *Taxodium distichum* after dry open storage for one winter.

Bates (1930) stored white pine seeds for one year in containers sealed with paraffin. He found that seeds stored at a low temperature, especially if they had been dried previously over sulphuric acid (30 per cent relative humidity), gave a higher germination percentage in less time from planting than did corresponding samples of fresh seed. The low moisture content was equally beneficial under all temperature conditions. Seeds of *Pinus longifolia* which had been sealed or mixed with charcoal in a gunny sack still gave good germination after two years (Champion, 1930). However, the best results were obtained when the seeds were well dried and sealed before the damp season. High moisture content and variable temperatures were found by Coile (1934) to cause a loss of vitality in slash pine (*Pinus caribaea* Morelet) seeds.

Isaac (1934) demonstrated that not only did cold storage preserve noble fir seed for five years without appreciable loss of viability, but it actually seemed to increase its germinating power. The same author (Isaac, 1935) showed that douglas fir seed, if left on the forest floor, either germinated or decayed within a year after it fell. Hence special storage is necessary. Baldwin (1934) reported the early loss of viability also of red spruce seeds when they were stored in duff, while during air-tight storage the germination decreased about 10 per cent each year for the first three years.

A detailed description of the results of storing some conifer seeds under controlled conditions has been given by Barton (1935*a*). The object of the experiments was to determine the effects of sealing, temperature, vacuum, and drying, on the keeping-quality of these seeds. Sealing was performed in glass distillation flasks. In case the air was to be exhausted from the flask, the top was sealed, after which the side tube was drawn out to a small bore and attached to an oil exhaust pump. When a good partial vacuum had been obtained, the side tube was sealed while still attached to the exhaust pump. In some tests, drying was accomplished by placing the seeds in desiccators over calcium oxide for sixty-one days prior to storage; in others, varying amounts of quicklime were mixed with the seeds before sealing. Storage was at room temperature, 5°C., and in a refrigeration room with an average temperature of −15°C. for the first two or three years of the experiments, after which the seeds were transferred to another refrigeration room with an average temperature of −5° to −4°C. Storage at room temperature was tried in the light as well as in the dark.

When viability tests were made, one seed sample was always planted directly in soil in the greenhouse while another sample was mixed with moist granulated peat-moss and placed at 5°C. Samples from this low-temperature treatment (stratification—*see* Glossary) were planted in the greenhouse after one or two, and, in a few cases, after three, months. In all cases, actual seedling production in soil was taken as an index of viability. Pretreatment at low temperatures for after-ripening was essential for some of the species studied. For example, in the tests of *Pinus taeda* L., if samples were planted directly in the greenhouse from the storage flask, one would conclude that the viability was lost after three years of storage, whereas with a pretreatment of one or two months in moist granulated peat-moss at 5°C., a high percentage of sound seedlings was produced. This is illustrated in Plate 5, where seedling production after storage for five years is depicted. *Pinus caribaea* Morelet and *Pinus ponderosa* Douglas showed the same marked effect of pretreatment at low temperature, while other forms studied, though less affected, benefited from the treatment.

The seedling production data from these experiments (Barton, 1935a) showed that sealed storage at low temperatures (5°C. or −4° to −15°C.) was effective for the maintenance of viability. *Pinus taeda* seeds under sealed, low-temperature storage retained their seedling-producing power fully for seven years, whereas in open room-temperature storage there was a decided decline in vitality after one year, and only a few seedlings were obtained thereafter. As compared with *Pinus taeda*, *P. caribaea*, *P. echinata* Mill., and *P. resinosa* Ait. all 'kept' only slightly better in open storage at room temperature and exhibited the same beneficial effects of sealing at low temperatures. *Pinus palustris* Mill. seeds lost viability much more rapidly. Seeds from open storage at room temperature lost their germination power completely in one year. Even under sealed, low-temperature storage there was a gradual decrease in seedling production to 50 per cent of the original after five years. *Pinus ponderosa*, *Picea excelsa* Link., and *Picea glauca* (Moench) Voss (*P. canadensis* of authors), stored only in refrigeration rooms with temperatures of −4° to −15°C., all kept well for four to six years.

All seeds were thoroughly air-dried before the experiments were started. Artificial desiccation, whether moderate or excessive, was found ineffective or harmful. A vacuum proved favourable to retention of

TABLE X

SEEDLING PRODUCTION IN THE GREENHOUSE FROM PINE SEEDS STORED AIR-DRY UNDER VARIOUS CONDITIONS

Species*	Storage Temp., °C	Open (O) or sealed (S)	Planted directly from storage								Pretreated at 5°C for 1 month							
			8	9	10	11	12	13	15	16	8	9	10	11	12	13	15	16
P. echinata	Lab.	S-vac.	34	14	14	10	9	8	10	6	25	10	18	17	19	20	0	0
	5	O	23	11	17	—	1	0	1	0	53	14	14	—	6	3	1	3
		S-air	8	17	31	—	15	8	18	10	35	16	39	—	50	38	36	16
		S-vac.	45	44	60	—	45	40	50	48	72	53	67	—	78	54	61	36
	−4	O	34	32	63	23	29	28	26	37	55	60	64	82	78	74	64	36
		S-air	24	33	27	11	27	21	8	27	41	16	39	48	41	49	35	42
		S-vac.	52	51	54	39	53	32	49	30	85	77	83	67	84	84	73	68
P. taeda	Lab.	S-vac.	16	12	27	15	5	5	13	—	35	34	30	19	33	43	22	—
	5	O	24	4	19	—	1	0	1	—	57	42	49	—	15	17	3	0
		S-air	18	3	17	—	7	3	8	1	49	25	41	—	13	28	15	3
		S-vac.	31	8	22	—	19	6	13	—	73	58	74	—	69	68	71	—
	−4	O	25	9	19	12	—	—	—	—	90	70	88	76	80	—	—	—
		S-air	45	4	11	3	12	1	—	—	78	48	69	51	73	74	—	—
		S-vac.	55	1	21	2	4	4	—	—	87	66	89	74	83	87	89	—

* At the beginning of the storage experiment, *P. echinata* and *P. taeda* gave up to 84 per cent seedling production in the greenhouse.
— Indicates that no test was made.

TABLE XI

SEEDLING PRODUCTION IN THE GREENHOUSE FROM CONIFER SEEDS STORED IN SEALED CONTAINERS AT APPROXIMATELY −4°C.

Per cent seedling production after storage for years

Species*	Wt. of seeds + CaO, gm.	Planted directly from storage							Pretreated at 5°C. for 1 month						
		8	9	10	11	12	14	15	8	9	10	11	12	14	15
Pinus taeda	100+ 0	0	1	2	0	0	—	—	44	40	46	64	55	—	—
	100+10	1	1	2	0	0	—	—	50	31	40	57	38	59	—
	100+17	0	0	0	1	0	3	1	53	30	29	66	39	51	45
	100+25	0	2	0	0	2	15	0	64	36	27	53	50	58	34
	100+32	0	1	0	0	5	3	—	48	21	20	52	46	41	—
	100+45	0	0	0	0	0	0	—	14	5	2	0	1	0	—
Picea excelsa	50+ 0	51	57	37	48	43	44	40	56	33	51	41	45	14	9
	50+ 8	33	29	23	24	27	23	37	21	37	14	17	16	2	0
Picea glauca	25+ 0	42	58	36	43	61	40	33	63	74	79	80	84	34	40
	25+ 4	34	19	15	17	38	18	1	70	53	63	63	67	12	8
	25+ 8	7	1	0	0	3	0	0	19	9	3	6	4	0	0
	25+15	3	2	1	1	4	0	0	20	11	4	7	12	1	0

* At the beginning of the storage experiment *Pinus taeda*, *Picea excelsa*, and *Picea glauca* gave up to 52, 65, and 48 per cent seedling production, respectively.
— Indicates that no test was made.

F

viability when the seeds were stored at room temperature, and both vacuum and low-temperature storage overcame in part the injurious effects of drying.

Some of these conifer seeds were tested further, after various periods of storage, until they were no longer capable of germination or until the seed supply was exhausted (Barton, 1953a). The data in Tables X and XI will suffice to show the trend of results secured from all of these seeds. A few effects are outstanding. Of the three temperatures tried, that of the laboratory resulted in more rapid deterioration than 5°C., which latter temperature was, in turn, less favourable than −4°C. for seeds of *P. echinata* and *P. taeda* (Table X).

Seeds stored in reduced air pressure obtained by means of a vacuum pump (vac.) kept better than those sealed in air or left without sealing. This was especially to be seen in laboratory storage where some seeds in evacuated flasks usually maintained viability for up to at least fifteen years of storage. This is in contrast to open or air-sealed storage at this temperature, where very few seeds survived for even two or three years. The data in Table X also point to the advantage of pretreating these pine seeds in moist granulated peat-moss for one month prior to planting in soil in the greenhouse—an effect already noted above. This is especially marked for *P. taeda* seeds.

Mixing pine seeds with different amounts of calcium oxide to attain different moisture contents for storage has the advantage of being a very simple and effective procedure, although care must be taken to avoid excess drying. Seeds of *P. taeda* so stored for eight to fifteen years in a room at approximately −4°C. showed the definite ill-effects of too much drying (Table XI). Again, the necessity for pretreatment of the seeds for seedling production in the soil in the greenhouse is demonstrated. Seeds of *P. palustris* mixed with small amounts of calcium oxide retained their germination capacity for nine years at −4°C.

Picea excelsa and *P. glauca* can be kept in sealed containers at approximately −4°C. for as long as fifteen years provided they are not dried too much (Table XI). It will be noted that the germination of neither of these forms was dependent upon pretreatment in a moist medium at 5°C. On the contrary, such pretreatment tended to decrease germination in some cases.

The efficacy of below-freezing temperatures for storage of conifer seeds has not been generally known until recently. A statistical analysis by Wakeley (1945) of Barton's 1935 data on pine seeds showed below-freezing temperatures to be significantly better for storage than 5°C. Results of subsequent tests on these seeds and many others have confirmed and extended this conclusion, as indicated elsewhere in this book (Chapter IV). From 2° to 4°C. has been recommended for all coniferous seed by Heit & Eliason (1940). Noble fir seed can be held for five years at 15°F. (Isaac, 1934), and a temperature of 32° to 39°F. is recommended for white pine seed storage (Roe, 1948).

Many other workers have studied the effects of storage conditions and ageing on the germination capacity of coniferous seeds, and for further information on particular species the reader is referred to the following: Anon., 1941; Barner & Dalskov, 1955; Crossley, 1955; Curtis, 1955; Göksin, 1942; Huss, 1954; Kjaer, 1950; Kondo & Kasahara, 1944; Lipkin, 1927; Mirov, 1946; M. L. Nelson, 1940; Rehackova, 1954; Schubert, 1954, 1955; Shirasawa & Koyama, 1915; Toumey, 1921; Vincent, 1929, 1937, 1937a, 1938; Wilson, 1953. Additional information may be obtained from the book on forest tree seeds by Baldwin (1942). Dent (1947) has written a seed manual for Indian forest plants, in which storage effects are included.

ELMS

Seeds of the American elm (*Ulmus americana* L.) belong to a group which normally live only a short time in open-air storage. It has been commonly supposed that these seeds must be planted as soon as they mature in early summer, with consequent impairment of seedling stands by the drought and heat of summer and, later, the necessity of protecting the tender young plants with the beginning of cold weather in the fall.

C. E. Steinbauer & Steinbauer (1932) desiccated elm seeds over different concentrations of sulphuric acid for a period of two weeks, after which they placed them in sealed containers at 0°, 10°, and 20°C. They found that low temperature and low water-content were favourable for the retention of vitality, the seeds remaining 'good' for up to nine months of storage but degenerating rapidly thereafter. Germination was much reduced after eleven months, which is the approximate time that fruits would have to be stored in order to be available for nursery planting in many localities. The germination tests of these authors were made between moistened blotters at 20°C. No soil tests were reported. They also noted that elm seeds have a resting period which is regardless of most storage conditions but is influenced both by the moisture content of the seeds and by storage temperature. No actual moisture determinations were made.

Delayed germination in American elm seeds has also been reported by Rudolf (1937), who obtained new seedlings in May 1937, from seeds planted in June 1936. He attributed this delay to extraordinary meteorological factors which also made possible the retention of vitality for one year.

George (1937) stored American elm seeds in containers with tight-fitting covers placed in an attic of an unheated building. Out-door temperatures ranged from 115° to −28°F. during the period of the test. As he obtained 79 per cent germination on moist blotters at room temperature after ten-and-a-half months of storage, he concluded that no special storage condition was necessary. Soil tests of these seeds yielded 33 per cent seedling production.

Manaresi (1935) found that the germinative capacity of the English elm (*Ulmus campestris* L.) remained fairly constant for two months and then diminished so that practically all of the seeds were dead after six months. He also stated that many of the fresh seeds did not germinate promptly but were more or less retarded up to a maximum of twenty-two days. This retardation, he said, depended on the 'embryonic post-maturation'.

Some of the results obtained by these various workers strongly suggested that measurement of the viability of elm seeds could be influenced markedly by the methods used for germination. This was confirmed by Barton (1939a), who tested the germination of fresh seeds of the American elm, *Ulmus americana* L., under a variety of conditions. As a result the following testing procedures were used to determine the germination capacity of American elm seeds after storage. At each testing period, fruits were mixed with moist granulated peat-moss and placed at controlled daily alternation of temperature of 10° to 25°C. Germination counts were made when the radicle had appeared and had begun to elongate. In addition to the oven tests, fruits were planted in soil in the greenhouse, using a mixture of equal parts of sand, sod soil, and granulated peat-moss. Fruit samples were planted directly from the storage containers. Other samples were placed in bottles in water and allowed to soak for twenty-four hours exposed to light, the source of which was a sixty-watt Mazda lamp placed at a distance of one foot above the bottles. After soaking, the water was drained off and the seeds were planted in soil. Still other samples were mixed with moist granulated peat-moss and placed at a constant temperature of 5°C. for one month prior to planting in the green-house. This method was shown to improve the seedling production markedly, although elm seeds are not commonly supposed to require low-temperature pretreatment. Seedling production was recorded when the seedlings had appeared above the soil.

These methods were used to test the viability of American elm seeds stored for various periods at room temperature, 5°C., or −5°C., in open containers or in sealed containers with air, oxygen, or a partial vacuum. Seed moisture contents of 7 (air-dry), 3, and 2 per cent were used. Results after sixteen months of storage are given in a paper published in 1939 (Barton, 1939a). The comparative germination capacities of fruits with 8 per cent moisture stored in open containers and in sealed tubes containing air, together with storage temperature effects, are shown in Fig. 10. The advantage of sealing and of low temperature is apparent. It was further shown that seeds with a reduced moisture content were superior in germination capacity after storage in sealed containers at room temperature for sixteen months. An atmosphere of oxygen proved especially deleterious when both moisture content and storage temperature were relatively high. A vacuum served to prolong viability in cases in which other storage conditions were unfavourable.

84

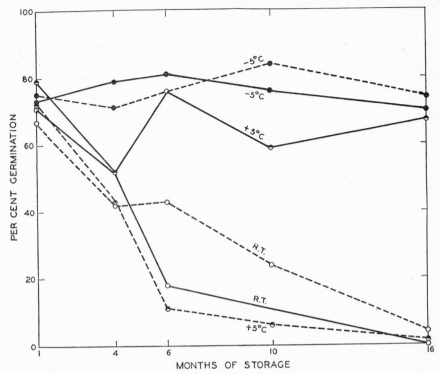

FIG. 10.—Graph indicating effect of storage of American elm seeds under various conditions upon germination at 10° to 25°C. daily alternation. Dotted lines indicate open storage; solid lines indicate sealed storage. R.T. = room temperature.

Up to 1939, sixteen months was the maximum life-span reported for American elm seeds. L. P. V. Johnson (1946) obtained some germination after one year of storage. Barton (1953a), continuing the tests on the seeds described above, found some of them still capable of germination after fifteen years of storage, as is shown in Table XII. Again in this table is shown the necessity for pretreatment of stored seeds to obtain a true index of their germination capacity. A comparison of the seedling production obtained in the greenhouse without pretreatment and after pretreatment in moist granulated peat-moss at 5°C. for one month makes this clear. Air-dry seeds, i.e. containing 7 per cent moisture at the time of storage, were stored in open as well as in sealed containers in the laboratory and at 5°C., but are not shown in the table because they did not survive two years of such storage. Seeds in open containers at —4°C. fared much better, germinating up to 51 per cent after six years of storage. Seeds dried to 3 or 2 per cent moisture remained viable longer than air-dry ones, regardless of temperature. As to the effect of temperature itself, there is no doubt as to the superiority of —4°C. over laboratory or 5°C. for maintaining viability. Safe storage under laboratory conditions did not exceed two years. At 5°C. the seeds could be stored for at least six years, and at −4°C. the life-span was extended to fifteen years. This extension of the life of elm seeds by controlled storage conditions is particularly noteworthy in view of their reported short life-span.

TABLE XII

SEEDLING PRODUCTION IN THE GREENHOUSE FROM AMERICAN ELM SEEDS STORED UNDER VARIOUS CONDITIONS

Germination pretreatment	Storage Temp., °C	Moisture content per cent	Container	Per cent seedling production after storage for No. of years stated 2	4	6	8	10	12	15
None	Lab.	3	Sealed	12	3	0	—	—	—	—
		2	Sealed	13	1	0	—	—	—	—
	5	7	Sealed	17	11	19	10	0	0	—
		3	Sealed	29	14	24	14	5	0	—
		2	Sealed	16	12	9	23	8	0	—
	−4	—	Open	10	14	4	10	4	0	—
		7	Sealed	14	22	15	16	26	15	7
		3	Sealed	22	40	22	33	29	13	12
		2	Sealed	17	43	18	18	22	18	—
1 month at 5°C.	Lab.	7	Sealed	70	2	—	—	—	—	—
		3	Sealed	55	33	0	—	—	—	—
		2	Sealed	58	11	0	—	—	—	—
	5	7	Sealed	90	75	78	15	2	0	—
		3	Sealed	82	84	73	51	25	1	—
		2	Sealed	82	83	55	31	18	0	—
	−4	—	Open	78	58	51	15	14	1	—
		7	Sealed	91	70	78	44	67	59	45
		3	Sealed	87	57	73	83	62	83	75
		2	Sealed	80	46	80	64	68	72	—

— Indicates that no test was made.

ASHES

Fraxinus (ash) seeds, also, are among those which have been reported to lose viability quickly under ordinary storage conditions. G. P. Steinbauer (1937), as a result of his experiments with *Fraxinus*, concluded that these seeds are much more sensitive to moisture content than to temperature—at least, over a period of a year in storage. Taylor (1941) reported a serious loss of viability of *Fraxinus* seeds by the end of one-and-a-half years of storage at warehouse temperatures. However, Barton (1945*a*) found that seeds of *Fraxinus* with moisture contents of approximately 7 to 10 per cent may be kept for at least seven years by using sealed containers at 5°C.

WILLOWS

Seeds of willows (*Salix* spp.) also lose their viability quickly when exposed to the air. This has been assumed to be due to excessive drying, but Nakajima's work disproves this assumption. He (Nakajima, 1921) found that seeds of *Salix opaca*, *S. japonica*, and *S. reinii*, retained their viability much better in closed tubes over a solution of 50 per cent by volume of sulphuric acid in water than they did in the open air. In later work he reported that seeds of *Salix pierotii* and *S. japonica* in the open air lost their ability to germinate within a week, but when enclosed over the sulphuric acid solution as mentioned above and stored at a low temperature, they still gave 53 per cent germination after 360 days of storage (Nakajima, 1926). Such a solution gives a relative humidity of only 13 per cent, which is much lower than the average humidity of the atmosphere at the ripening time of the seeds. Evidently the injury in the open air is not due to excessive drying. The life of *Salix caprea* seeds, which is normally only thirty to forty days, can also be extended by drying the seeds and storing them at a temperature of 6° to 9°C. (Janiševskii & Pervuhina, 1941).

BIRCHES

Seeds of *Betula lenta* and *B. populifolia* 'kept' perfectly during one year of air-dry storage at room temperature, while those of *B. papyrifera* suffered loss in viability (Joseph, 1929). The optimum moisture content for seeds stored at room temperature in sealed containers lies considerably below that of freshly harvested seeds for *B. papyrifera* (0.6 per cent), while *B. populifolia* seed keeps best with a medium amount of hygroscopic moisture (5·2 per cent). *B. lenta* seed keeps well for one year in all except very humid conditions. The moisture content proved to be unimportant during one year's storage of the three species of *Betula* in a refrigerator.

POPLARS, ETC.

Engstrom (1948) recommends that cottonwood (*Populus* sp.) seeds be planted immediately after harvest, since there is much loss of viability within two days. Seeds of American aspen (*Populus tremuloides*) and balsam poplar (*P. balsamifera*), on the other hand, gave vigorous germination at the end of four weeks, and 45 per cent germination at the end of eight weeks of storage in open dishes (Moss, 1938). After eight weeks, however, deterioration was rapid. Although these seeds have been reported not to tolerate desiccation, storing over calcium chloride at −5°C. showed remarkable extension of life, seed of the Russian poplar (*P. petrowskyana*) giving 70 per cent germination after two years. Storage of *Populus tremuloides* and *P. grandidentata* seeds at room temperature in a controlled-humidity series showed the optimum relative humidity to be 20 per cent, where seeds of the former remained viable for 555 and those of the latter for 455 days, as compared with a life-span of only 28 days in the open air (L. P. V. Johnson, 1946). Busse (1935) presented evidence to show that the loss of viability of aspen seeds is due to the injurious action of oxygen; also that higher temperatures hasten the degeneration.

SOUTHERN-BEECHES

Nothofagus menziesii seeds stored in the open air will not remain viable over winter, but in sealed containers at 2° to 5°C. full germination was retained (Bibby & Williams, 1953).

FRUIT TREES

The ability of rosaceous seeds to withstand dry storage is of great practical interest, because, in the case of species used for understock—such as *Prunus americana* Marsh—heavy crops sometimes appear only in alternate years, which makes it desirable to carry over seeds for poor-bearing years. Also, with rosaceous seeds that have a short after-ripening period, it is convenient to keep the seeds in dry storage until the time for stratification in mid- or late winter. In 1851, Fleischer stated that rosaceous seeds retain their viability in dry storage for considerable periods (Fleischer, 1851). Seeds taken from apples and stored dry for two-and-a-half years germinate almost as well and give nearly as high a seedling production in soil as do seeds one-and-a-half years old of the same sort stored similarly (Crocker, 1928; Crocker & Barton, 1931). However, both of these show considerably lower germination percentages than do seeds only half-a-year old. Apple seeds that are to be used to produce under-stock, or hybrid-apple seeds, will keep well if they are removed from the apples in the autumn, cleaned, dried, and stored until it is time to stratify in late winter for spring planting.

Seeds of the American wild plum (*Prunus americana* Marsh) retain more than 60 per cent of their power to produce seedlings after fifty-three months of dry storage at 7° to 10°C. (Giersbach & Crocker, 1932). Optimum germination of apple, pear, apricot, cherry, plum, and sour cherry seeds occurred after storage at a relative humidity of 50 to 55 per cent (Solovjeva, 1950). Higher humidities were detrimental. Storing peach seeds dry for one year decreased their value (Brase, 1948).

It appears that no long-term storage investigation under a variety of conditions has been attempted for fruit-tree seeds. Such an experiment would undoubtedly reveal methods of extending the life-span of these seeds.

A number of other tree seeds have been the subject of longevity studies. *Eucalyptus miniata* seeds collected in 1867 gave excellent germination in 1880 (Müller, 1880). Successful storage of a number of Hawaiian forest tree seeds depended upon lowering of the relative humidity and the temperature (Akamine, 1951). These same conditions preserved the quality and the germination capacity of pecans (C. L. Smith *et al.*, 1933; Wright, 1941; Brison, 1945).

QUININE

The keeping-quality of *Cinchona* seeds under various conditions is a matter of importance in planting and breeding programmes of this genus. It is necessary to keep a viable supply of seeds for continuous nursery propagation in order to supply seedlings for field plantings of the trees. Also, where extensive hybridization is practised, seed storage methods, which permit retention of germination capacity of the hybrid seeds until the desirability of the trees produced from a sample planting can be determined, are essential. This might mean a storage period of from four to ten years. It has been reported that seeds older than 186 days at the time of planting in Puerto Rico germinated poorly (Anon., 1940*a*). In Uganda, A. S. Thomas (1946) secured 90 to 98 per cent germination from fresh seeds, 70 to 80 per cent from seeds nine months old, 50 per cent from seeds twenty months old, and only 5 per cent after the seeds had been stored for thirty-two months. Morrison (1944) stated that *Cinchona* seeds lost about 5 per cent of their germination power in about two years. Storage was at room temperature in closed containers with calcium chloride or sulphuric acid as a drying agent. No data were given. Air-tight containers with calcium chloride also prolonged the viability of *C. ledgeriana* Moens. in Puerto Rico (Kevorkian, 1941). A refrigerator as well as room temperature was used in the last instance, but little temperature effect was noted up to twelve months of storage, when 33·5 and 30 per cent germinations were obtained, respectively.

In an article on *Cinchona* seed storage, Kreyer (1939) recorded information received from the Biological-Agronomical Institute in Tanganyika to the effect that the germination capacity of freshly harvested *Cinchona* seeds reached 99 per cent. Five months after gathering, however, the viability had become reduced by 4 per cent, and after eleven to fourteen months only about 50 per cent of the seeds were capable of germination. Kreyer stored seeds of *C. ledgeriana* for 418 days with various drying agents in hermetically closed jars at a temperature of 25 °C. Seeds showed a high germination capacity when kept in jars with large doses of sodium bromide, calcium chloride, a special desiccant called 'Adorosu', or an average dose of caustic lime. None of the controls germinated after 418 days.

89

Cowgill (1944) has done extensive work on *Cinchona*, part of which dealt with seed storage for fifty-four weeks but with variable results.

Data are available to show the effects of temperature, moisture content, light, and gaseous exchange, upon the keeping-quality of seeds of *Cinchona ledgeriana* for periods up to and including four years (Barton, 1947). Seeds with 4·8, 6·0, and 9·4 per cent of moisture were stored in open and sealed containers in the laboratory, at 5°C., and at approximately −4°C. Deterioration in the laboratory was evident after eighteen months, and germination was reduced to approximately one-half or less of the original capacity after twenty-four months. There was no advantage of sealed over open storage at this temperature. When the moisture content of the seeds was as high as 9·4 per cent, there was some evidence of harmful effects of sealing and of light in the laboratory. Loss of germination capacity also proceeded rapidly in open storage in a humid, 5°C. room. On the other hand, seeds stored sealed at 5°C., or open or sealed at approximately −4°C., kept perfectly during up to forty-eight months of storage. Within the limits of this experiment, temperature appeared to be a more important factor than moisture content in the effects on *Cinchona* seeds. It was concluded that *C. ledgeriana* seeds can be kept successfully for at least four years by maintaining a moisture content of below 9 per cent of the dry-weight of the seeds and keeping the temperature between 5°C. and −4°C. (Barton 1947).

Tree seeds which will not tolerate drying will be discussed in the next chapter.

X

LONGEVITY OF SEEDS OF AQUATIC PLANTS, ETC.

IN a consideration of the storage in water of the seeds of aquatic plants, two things must be kept in mind. First, only dormant seeds can be stored in water at temperatures permitting germination, and second, many seeds of water-plants of the temperate zone after-ripen in water at temperatures near or several degrees above the freezing-point. When such seeds are fully after-ripened, their germination may take place at the low temperature, and will certainly occur upon transfer to a higher temperature—provided, of course, that other conditions are suitable and the seed is viable. Long preservation of aquatic seeds in water, therefore, is dependent on the dormancy of the seeds and persistence of a temperature which will not break the dormancy. The behaviour of *Alisma plantago-aquatica* fruits illustrates both these points. If the intact fruits or seeds are placed in water and kept at room temperature, they will remain intact and viable for months or even years on end with little or no germination. If transferred in water to any temperature between 1° and 10°C. for one-and-a-half months, they after-ripen and germinate freely when returned to room temperature. They begin germinating even at the after-ripening temperatures if left there for four months. These seeds also germinate readily at room temperature if frozen and thawed repeatedly while full of imbibed water, or if the seed-coats are broken mechanically. Freshly harvested or dry-stored seeds behave in the same manner as those stored at room temperature in water.

Isely (1944) noted the dormancy of fourteen species of *Scirpus* seeds at maturity, and the need for an after-ripening period during which the seeds should be kept in a moist medium or in water at 2° to 4°C. Many species of *Scirpus* appeared to have an extended period of viability when properly stored.

Muenscher (1936, 1936a) stored seeds of thirty genera and sixty-one species of temperate-zone water-plants under four different conditions: in water at 1° to 3°C., in water at the laboratory temperature, and in air at these two temperatures. Germination tests were made in water in a greenhouse after about two, five, and seven months of storage in each condition. Seeds of the following plants showed little or no germination, except after storage in water at 1° to 3°C.: fourteen species of *Potamogeton*, three species of *Najas*, *Alisma plantago-aquatica*, three species of *Sagittaria*, *Butomus umbellatus*, *Zizania aquatica*, two species of *Scirpus*,

91

Acorus calamus, Calla palustris, Orontium aquaticum, Peltandra virginica, Heteranthera dubia, Pontederia cordata, Eriocaulon septangulare, Polygonum amphibium, Ceratophyllum demersum, Nymphozanthus variegatus, Nymphaea tuberosa, Trapa natans, Elatine americana, and *Lobelia dortmanna.* The percentage of germination generally increased with the period of cold storage from two to five and finally to seven months. No doubt some or all of these seeds were dormant and were after-ripened by storage in water at 1° to 3°C. This would account for the increased germination after the longer periods in cold water. It seems that even seven months of storage was not sufficient to produce full after-ripening in some of the seeds.

Seeds of the following species showed little or no germination after any of the conditions of storage used by Muenscher: five species of *Sparganium, Najas minor, Sagittaria latifolia, Lophotocarpus spongiosus,* three species of *Scirpus,* and *Eleocharis calva.* Whether this was due to poor seed quality or a failure to provide proper germination conditions is not evident from the data.

Muenscher found that *Glyceria striata* and *Nasturtium nasturtium-aquaticum* showed fair to good germination after all four conditions of storage. *Vallisneria americana* germinated readily in water at the laboratory temperature, whether it had previously been stored in cold water or not. Of the species tried, only one germinated after dry storage.

Guppy (1897) gives the records of the behaviour of the seeds of a number of aquatic plants when stored in air and in water. Seeds of *Nuphar luteum* and *Nymphaea alba* are soon killed by drying. Seeds of the former germinate more readily than those of the latter, but some are delayed until the second year or later. *Potamogeton* seeds are also injured by protracted drying. However, seeds of the following water-plants endure drying without injury for the period mentioned after each, and this is the longest period determined by Guppy unless otherwise noted: *Myriophyllum spicatum,* $1\frac{1}{2}$ years; *M. alterniflorum,* 1 year; *Limnanthemum nymphaeoides,* $2\frac{1}{2}$ years; *Sparganium ramosum, S. simplex,* and *Zannichellia palustris,* $4\frac{1}{2}$ months; *Callitriche aquatica,* 2 years, but rotted after drying $3\frac{1}{2}$ years; and *Ranunculus aquatilis* and *R. hederaceus,* many months. Drying for two months promoted the germination of *Ceratophyllum demersum,* and drying for three or four months increased the germination of *Sagittaria sagittifolia* seeds.

Seeds of *Alisma plantago-aquatica, Typha latifolia,* and *Butomus umbellatus,* stored dry in envelopes in the laboratory at Yonkers, New York, were tested for germination by breaking the fruit and seed-coats and placing in water at a daily alternating temperature of 15° to 30°C. *Alisma* $12\frac{1}{2}$, $6\frac{1}{2}$, $3\frac{1}{2}$, and $2\frac{1}{2}$ years old gave 0, 40, 39, and 51 per cent germination, respectively; *Typha* $12\frac{1}{2}$, $5\frac{1}{2}$, and $4\frac{1}{2}$ years old gave 0, 96, and 78 per cent germination; and *Butomus* $7\frac{1}{2}$ and $3\frac{1}{2}$ years old gave 9 and 26 per cent germination.

It has been shown by Haigh (1940) that seeds of the water hyacinth (*Eichhornia crassipes* Solms.) remain viable in water in the laboratory for at least five years, but that dry seeds fail to germinate after three years of storage. Seeds of this plant were able to survive one month in ice at approximately −4°C., and as long as two months at temperatures up to 15°C. (Barton & Hotchkiss, 1951). At 40°C., also, the seeds not only survived for two months, but gave excellent germination after such storage. When the storage period was lengthened to seventeen months, however, 20° and 30°C. proved better than 5° or 40°C. for keeping the seeds viable. The life-span of the seeds under these various conditions was not determined, but the requirement of a combination of high temperature and light for complete germination was established. Hitchcock *et al.* (1949, 1950) kept water hyacinth seeds in water in the laboratory for more than two years without germination. A knowledge of the germination behaviour and the life-span of seeds of *Eichhornia* is important in the control of this plant in waterways, as enough seedlings may become established in a year to pose a serious threat of reinfestation following successful control by chemicals of the vegetative parts of the plants.

Longevity is also an important factor in a control programme for *Halogeton glomeratus*, the seeds of which decline in viability after one year (Anon., 1953).

Certain species of *Polygonum* can live either in water or on land, and are important as food and cover for water-fowl and fish, but, under some conditions, these same plants may become tiresome weeds. Viability and dormancy in three *Polygonum* species, *P. amphibium* L., *P. coccineum* Muhl., and *P. hydropiperoides* Michx., have been studied by Justice (1944). Achenes of all three species placed under outdoor conditions similar to their habitats lost their germination capacity during the first winter. Also, seeds of all three were killed by a year of dry storage in the laboratory. They could be stored successfully for longer periods in water at 2°C., where any dormancy present was broken.

Duvel (1905) found that wild rice (*Zizania aquatica*) seeds lost their germination capacity if they were allowed to dry in the air for even a few days, but that they retained their viability perfectly until spring if stored in water at 0° to 1°C. In the spring they must be transferred from the storage water to the water in which they are to grow, without being allowed to dry. Seeds of wild rice are dormant when mature, and storage in water near the freezing-point will after-ripen them while maintaining their viability.

The quality of cultivated rice (*Oryza sativa* L.) grain stored under a variety of conditions was reported on by Kondo and his co-workers in a series of twenty articles published over the period 1927–38. The proper amount of drying is essential to the maintenance of satisfactory food-qualities as well as viability. Drying by the use of calcium chloride (Kondo

93

& Okamura, 1931, 1934*b*, 1934*c*, 1935; McFarlane *et al.*, 1955) or calcium oxide (Kondo & Isshiki, 1936; Kondo & Terasaka, 1936) is recommended, though heated air can also be used (Kondo & Okamura, 1932–33, 1932–33*a*; Kondo *et al.*, 1938). A more modern method is pre-harvest drying with chemicals sprayed from an airplane (Anon. 1953*a*). The safe moisture content of rice grain depends upon the storage temperature (Kondo & Okamura, 1938). Rice seeds stored for five years in a concrete silo, hermetically sealed (Kondo & Okamura, 1937), or for twenty-three years on a mountain (Kondo & Okamura, 1934*b*), retained satisfactory culinary qualities, but were not capable of germination. Unhulled rice stored for about 100 years in a granary was still edible, though it had an unpleasant taste (Kondo & Okamura, 1938*a*).

Sampietro (1931) was able to keep rice seeds fully viable (99 per cent germination) for eight years provided they were dried to a moisture content of 5 per cent and stored in an atmosphere of nitrogen. In atmospheres of carbon dioxide or air, or in a partial vacuum, the seeds were killed. Thirteen per cent moisture was fatal under all conditions. Similar results were obtained by Saran (1945). Dore (1955) stored rice grain in an air-conditioned room of low temperature and humidity, where the seeds retained their viability for three-and-a-half years. They remained viable for two years in sealed containers with calcium chloride.

We have seen that seeds of some water plants will not survive storage in the air, while others will stand several years of dry storage in the air without injury, and *Nelumbo* (*see* Chapter I) seeds live for some centuries in dry storage.

Seeds Injured by Desiccation

Some seeds lose their germination capacity in a very short time when they are kept in the open air after harvest. This has been assumed to be due to the drying effect of the air. As we have seen in Chapter IX, many seeds, such as those for example of the American elm and some aquatic plants, which are supposed to be killed by drying, tolerate desiccation to a remarkable degree and can be kept for long periods in a dry state at low temperatures. However, there remain certain seeds which appear to be killed by drying. Though most of these are of tropical or semi-tropical origin, there are some temperate-zone forms among them.

According to H. A. Jones (1920), the seeds of the river maple (*Acer saccharinum*) are killed by relatively slight drying. When they fall from the tree in June, they bear about 58 per cent of water. Regardless of the temperature of exposure (0° to 35°C.), they were killed when the moisture content reached 30 to 34 per cent. In Jones's experiments it required six days at 35°C. and ninety-two days at 0°C. to reach this water content or the death point. When these seeds were stored in a closed vessel over water at the freezing-point and provision was made for preventing carbon dioxide accumulation, they retained full viability for 102 days, which was the limit of the test. The low temperature prevented germination and reduced the rate of metabolism. The latter is an important consideration in any seed with high moisture content. River maple seeds should be sown immediately after harvest. If this is impossible, because of the necessity of shipping or for any other reason, they should be kept near the freezing-point, and water loss prevented. They are not dormant but begin germination in nature as soon as they reach the moist ground. The seeds of the fall-fruiting sugar maple (*A. saccharum*) show very different behaviour. They endure complete air-drying and respond to about three months' low-temperature stratification* for eliminating dormancy.

Acorns (the seeds of oaks, *Quercus* spp.) have a very short life-span under ordinary conditions of storage (Oppermann, 1913; Gardner, 1937), but extension of germination capacity up to ten months results from mixing with dry sand and storing in air-tight cans at 32° to 40°F. (Mirov, 1943). Seeds of the tulip-tree (*Liriodendron tulipifera*), which are injured by drying, have been kept for four years without loss of viability by storing in soil (Paton, 1945). Six thousand pounds of seed were placed in a pit in two-inch layers alternating with layers of sand to cover the seeds. A concavity to catch water was provided at the top of the pit, so that the seeds were kept moist.

Kidd (1914) attributed prolongation of the life of rapidly deteriorating *Hevea brasiliensis* seeds to the presence of 40 per cent of carbon dioxide produced by the respiration of the seeds in a closed flask. He did not make moisture determinations of these seeds but stated that they had a high water content. He summarized, in part, that 'the resting stage of the moist seed is primarily a phase of narcosis induced by the action of carbon dioxide' (Kidd, 1914, p. 624). This is contrary to the claim made by H. A. Jones (1920) that *Acer saccharinum* seeds kept better when carbon dioxide was not allowed to accumulate than when it was.

Busse (1935) thought that the rapid loss of viability of poplar seeds when left in air was due to the injurious action of oxygen. He was able to demonstrate that seeds in sealed containers with reduced oxygen pressure remained viable longer than those subjected to full atmospheric pressure.

Sugar-cane seeds degenerate rapidly when stored in the open air. This makes it impossible to ship them with assurance from one sugar region of the world to distant ones where seedlings are desired for breeding. Verret (1928) found that viability could be lengthened materially by taking the seeds from the thoroughly air-dried heads, placing them in cans with 9 gm. of calcium chloride to 1 litre of space, displacing the air with carbon dioxide, hermetically sealing, and storing at the freezing-point. In these seeds, low and perhaps constant water-content and absence

* 'Stratification' is a term derived from the practice of nurserymen to spread layers of seeds alternating with layers of moist soil or sand for over-winter or other low-temperature treatment of seeds. The same term is now generally used to denote any low-temperature treatment of seeds in a moist medium.

of oxygen seem to be necessary for retention of germination capacity. It is possible also that carbon dioxide may play a positive role rather than having its value in merely displacing oxygen. These results were confirmed in the main by Darragh (1931). More recently, freezing at 0°F. has been added to drying for successful storage of sugar-cane seeds (Abbott, 1950; Anon., 1951a, 1952).

Citrus seeds are among those which are injured by drying and hence deteriorate rapidly under ordinary conditions of storage. Dealers in these seeds have experienced difficulty in maintaining their germinative power, although Florida sour orange seeds have been reported to be more resistant to drying effects than most other citrus seeds. For the most part the seeds have been kept moist, either in the original fruit juice or in moist sand or some other moist medium after cleaning, until they were sold. Inquiries from seedsmen, together with the lack of experimental work on the keeping quality of these seeds, led to a study reported by Barton (1943a). Seeds of grapefruit, sweet orange, sour orange, and rough lemon, were stored under different conditions to determine factors which would prolong life.

Work with grapefruit seeds has shown the importance of moisture content, temperature, and sealing effects, to viability. High moistures (100 to 131 per cent calculated on the basis of dry-weights of the seeds) were detrimental to seeds in closed containers, especially during storage at laboratory temperatures. However, drying of grapefruit seeds to 60 per cent of the dry weight, which is still a very high moisture content, prolonged their viability in sealed storage in the laboratory as compared with open storage at this temperature. Obviously, then, some factor other than carbon dioxide accumulation was operating to prolong the life of grapefruit seeds, as sealed storage was not favourable in all cases. This was shown further by the less favourable effect of sealed storage at 5°C. as compared with open storage at that temperature. Mould appeared at an early date on moist seeds in sealed storage in the laboratory and may have been a factor in deterioration, but it would be very difficult to say whether the mould led to deterioration or whether deterioration led to the development of mould.

The best storage condition found for grapefruit seeds was in open containers in a room held at 5°C. After a year of open storage in this very humid 5°C room, grapefruit seeds contained 18 per cent of moisture. As limited tests indicated that fresh seeds were somewhat injured by drying on blotters in the laboratory until they contained 52 per cent of moisture, the beneficial effect of drying to 18 per cent at 5°C. is of particular interest. Certainly it cannot be said that their longevity depends entirely upon water content. It appears that the drying temperature and perhaps other factors such as rapidity of desiccation, etc., are of importance. In this respect, these seeds differ from those of river maple as described by H. A. Jones (1920). Citrus seeds apparently vary widely in their tolerance to desiccation. Whereas the germination of seeds of grapefruit was reduced 30 per cent by drying to 52 per cent of moisture, and that of sweet orange seeds was reduced 80 per cent by drying to 25 per cent of moisture, sour orange and rough lemon seeds could be dried to 4 per cent of moisture and still retain 25 and 50 per cent of their original germination capacity.

The harmful effect of a temperature of approximately $-5°C$. on the keeping quality of seeds of grapefruit, sour orange, and rough lemon, is in direct contrast to the efficacy of this same temperature for maintaining the viability of many seeds which can be dried to relatively low moisture contents. That the deleterious effect on citrus seeds of such temperatures was not due entirely to high moisture content was demonstrated in the experiments of Barton (1943a), which showed that rough lemon and sour orange seeds dried to 4 per cent of moisture and still capable of germination, were unfavourably affected by below-freezing temperature. Childs & Hrnciar (1949) believed that micro-organisms play an important role in the deterioration of citrus seeds in storage. They stored large lots of thirty-four varieties for six months and of thirty-three varieties for eight months at 35°F. (1·7°C.) in moist sawdust or moist moss in unsealed containers, after dipping the seeds in a 1 per cent solution of 8-hydroxy-quinoline sulphate. Twenty-nine of the lots gave better than 90 per cent germination after six months and the average germination after eight months of storage was 84·5 per cent. There were, however, great differences between varieties. Deterioration of citrus seeds in storage was attributed to micro-organisms, moisture, temperature, and aeration. Richards (1952) found a marked reduction in viability of seeds of Country lime and Bibile sweet orange when they were stored at normal air temperatures of 78° to 80°F. for more than three days after extraction from the fruit. The interrelation of gaseous exchange, moisture content, temperature, and micro-organisms, in their effects on the keeping quality of short-lived citrus seeds, is not apparent at the present time. Much more work is needed to clarify the situation.

The rapid deterioration of seeds of maga (*Montezuma speciosissima* Moc. & Sessé) after harvest, coupled with the economic importance of the tree in yielding wood of good quality, has made it desirable to determine conditions for keeping seeds viable over longer periods of time than they usually live. Germination occurs shortly after the seeds fall from the tree, and may even start before the fall of the capsule. From this characteristic one would judge that desiccation might destroy the life of the embryo, which is large and occupies all of the space within the seed-coat. Seed storage of maga was the subject of a study made by Barton (1945). It was found that the life-span became reduced significantly within two weeks under ordinary storage conditions. When the moisture content was as high as 33 per cent of the wet weight, the seeds could be kept for a month in sealed storage at 5°C. Drying at the laboratory temperature to approximately 10 per cent moisture reduced the germination capacity immediately by one-half, but permitted subsequent retention of this lowered germination capacity at $-5°$, $5°$, or $20°$C. for at least six months, and possibly longer.

These results are in general agreement with those of Marrero (1942), who dried seed-lots of maga with the aid of an electric fan to 87·5, 75·0, and 62·5 per cent of the original weight.

Refrigeration is necessary to extend the life of many tropical hardwood seeds from two or three months to one year (Marrero, 1943). Seeds of the Cupuliferae are harmed by dry storage, but may be kept for as long as two years if they are mixed with river sand and held at 0° to 5°C. (Kondo *et al.*, 1941). The viability of air-dry seeds of mangosteen (*Garcinia mangostana* L.) is lost after four weeks' storage, but may be preserved for seven to eight weeks in moistened charcoal or peat at room temperature (Winters & Rodriguez-Colon, 1953). Storage for one week over calcium chloride killed all the seeds. Other tropical seeds must be kept moist to maintain viability, and thus do not have long life-spans in storage. Among these are seeds of cacao (Evans, 1950), coffee (González-Rios, 1929; Ultée, 1933; Anon., 1941*a*), *Erythroxylon* and *Litchi* (McClelland, 1944), large-leaf mahogany (Lopez, 1938), and *Macadamia* (Chu *et al.*, 1953).

D. G. White and his co-workers have investigated the storage requirements of bamboo seeds (D. G. White, 1947; D. G. White & Villafane, 1947; D. G. White & Delgado, 1948) and have found that the most practical method of preserving their viability is storage over calcium chloride at room temperature, though refrigeration results in still greater life-extension. Webster (1948) tried several methods of storing seeds of the tung (*Aleurites*) tree; but the only one which prevented loss of viability was stratification with moist sand followed by storage in boxes in a cool building. Under these conditions full germination capacity was retained for six-and-a-half months. The life of these seeds is greatly prolonged by storage at 34° to 40°F. according to Large *et al.* (1947), but Shear & Crane (1943) reported that tung seeds stored at 7°C. lost their viability sooner than did seeds from storage temperatures which ranged either from 23° to 32°C. or from 7° to 55°C.

It has been shown above that, although seeds of all kinds deteriorate very rapidly in the tropics, it is possible, through humidity and temperature control, to extend the life-span of many valuable seeds, thus permitting the correlation of direct seeding with correct planting schedules.

XI

EFFECT OF PACKETS ON SEED VIABILITY

It has been well established that moisture content and temperature determine, to a large extent, the keeping quality of seeds in storage. Neither moisture nor temperature can be considered alone, as the effect of one depends upon the other. These factors not only affect seeds while they are in controlled storage rooms maintained by seedsmen, but continue to exert a vital influence after seeds are removed and packeted for the trade. Furthermore, it is essential that the seeds maintain full viability under the original storage conditions, in order to withstand further storage under unfavourable conditions. The better the storage condition for keeping the seeds viable, the better the chance of survival when they are removed to an unfavourable environment. Other conditions being equal, low temperature has been found to be superior to higher temperatures for the maintenance of germination capacity. However, seeds dried and stored in sealed containers at ordinary room temperature may keep better than those stored open in a room at 5°C. where high humidity prevails, as has been demonstrated repeatedly in storage experiments at Yonkers, New York. In these cases, then, resistance to further storage would be in favour of the original room-temperature storage or the storage conditions permitting the highest retention of germination capacity.

The keeping quality of any particular lot of seeds does not depend upon its initial high quality, but upon the storage conditions. Seeds of low germination capacity may be stored successfully for fairly long periods if the storage is appropriate. On the other hand, seeds in which deterioration has been initiated, even if the germination capacity is still high, are incapable of remaining viable for long periods under adverse storage conditions.

All of these facts point to the necessity of maintaining rooms in which temperature and humidity can be controlled, if commercially important seeds are to be held over for sale in subsequent years. Responsible seedsmen have come to realize this and have provided the proper facilities. However, another problem, that of deterioration upon removal from storage, is being recognized as of the utmost significance. Seedsmen must packet their seeds for the trade well in advance of the time the consumer buys and plants them. Kraft paper envelopes are commonly used, and are sent to wholesale and retail dealers in many parts of the world. Some of these packeted seeds are stored in warehouses or retail stores under conditions of high temperature or humidity, or both, where germination capacity decreases at such a rapid rate that the seeds may be worthless when the customer finally plants them. The rapid deterioration under unfavourable conditions of onion seed upon removal from storage was reported by Barton (1939, 1941).

Some idea of the importance of the seed-packet industry and of the poor quality of seeds found in many such packets may be gained from a series of articles by M. T. Munn and his co-workers, published between 1924 and 1949. Two of these will serve as examples (Munn *et al.*, 1929; Munn & Munn, 1936). In 1926, 1927, and 1928, packets of vegetable seeds were purchased from forty-seven different seedsmen or dealers in New York. Out of 964 packets, 9·2 per cent were worthless for planting purposes. In addition to these, there was a relatively high percentage of packets of very little seeding value. In 1923, 23 per cent of 347 packets, and in 1924, 18 per cent of 600 packets, were worthless. This represented very serious losses to the growers. The analysis, by these workers, of the germination behaviour of the seeds—by firms, with an actual listing of the firms and the quality of seeds they sold—has doubtless contributed to the forcing of some unscrupulous dealers out of the market.

An analogous situation has been found for flower seeds in New York (Munn & Munn, 1936). In 1936, of 500 packets of flower seeds examined, 57 per cent were excellent, good, or entirely satisfactory, while the remaining 43 per cent were unsatisfactory. Fifteen per cent were worthless. Similar work has been reported from the Massachusetts Agricultural Experiment Station (Snyder & Tuttle, 1937), from the Seed Laboratory of the Department of Agriculture, Canada (Dymond, 1917), and the California State Department of Agriculture (Suskin, 1953).

As a result of these and probably many other unreported tests, there are now Federal and State seed laws in the United States which seek to govern the type of seed which is offered for sale. From the fundamental seed behaviour described in previous chapters, however, it is evident that the problem of control is dependent upon a basic knowledge of the response of seeds to storage conditions—not only in the special storage room of the responsible seedsman, but also from the time they leave that room until they reach the grower, and, ultimately, the soil.

Among the published reports on the germination of packeted seeds is that of E. Brown & Goss (1912). Over a period of five years (from 1907 to 1911) they tested 18,571 packets of vegetable seeds. These were the familiar paper packets lithographed in colours and put up by sixty different firms. These paper packets are shipped in boxes and hence are known as box seeds. Mail-order seeds from twenty firms were also included in the study. A marked difference in the quality of the box and mail-order seeds was noted. Whereas the average germination of all of the former was 60·5 per cent, that of the latter was 77·5 per cent. When the germination of box seeds by firms was analysed, the difference in average germination for the different firms was marked (from 81·5 to 36.5 per cent), while the difference in average germination of seeds from mail-order houses was small (from 81·8 to 70·3 per cent). Differences in box and mail-order seeds of the same variety ranged from 2·3 per cent for parsnip and 3·3 per cent for spinach, to 28·5 per cent for cabbage and 29·5 per cent for onion. The average germination of box seeds put out by any one firm was uniform from year to year, but the differences in quality varied from 73·7 per cent for one firm to 42·8 per cent for another. Many instances were found in which there were statements of high germination on the packet and yet the authors obtained only 42·8 to 43·6 per cent germination. They concluded (E. Brown & Goss, 1912, p. 9): 'It would thus appear that such firms either do not test their seeds accurately or else disregard the results of such tests to the extent of including lots showing unsatisfactory germination in their output of packeted seeds for the box trade.' The dependence of mail-order firms on satisfied customers was given by Brown & Goss as the reason for the superiority of their seeds.

In the light of the evidence now available, the better quality of the mail-order seeds might be attributed, at least in part, to the direct route from the favourable storage rooms of the seedsmen right to the consumer. Such seeds would tend to be subjected to possible high temperature and humidity conditions for a minimum period and hence have a higher germination capacity at planting time than the box seeds. Also, it may very well be that seeds which may have given 80 to 100 per cent germination at the time they were packeted for distribution have lost their value entirely before they reached the grower, if the wholesale or retail establishment has poor storage facilities. While this fact may remove some of the blame for seed failure from the seedsman, it is still his responsibility to provide the best possible seeds for filling the packets he sends out.

As the moisture content of the seeds is such an important factor in their deterioration, and as the paper packets commonly used in the trade are not in any sense waterproof, it became of interest to determine the difference in viability of seeds in these packets as compared with ones in waterproof packets. The results to be presented below have demonstrated the superiority of tin cans with tight-fitting lids, or of heat-sealed aluminium foil envelopes, over the ordinary kraft paper seed-envelope for the successful packeting of onion seeds (Barton, 1949).

Germination tests and moisture determinations were made immediately after receipt of seeds of onion (*Allium cepa* L. var. Ebenezer). The moisture content of two portions of the seeds was adjusted to 10·6 and 3·5 per cent of the dry-weight of the seeds by placing the seeds in desiccators over water or calcium oxide until the desired weight was obtained. The seeds were then stored at 5° and 25 °C. in open and in sealed containers.

Viability tests were made after 3, 6, 12, and 24 months of storage. At the end of each of these periods, seeds from each storage condition were packeted in paper seed-envelopes of the type usually employed in the trade, and in tin cans with tight-fitting lids sealed with sealing wax.

These packets and tin cans were then placed for further storage in a room at 30°C. The paper packets were stored in a desiccator over a saturated solution of sodium chloride giving approximately 76 per cent relative humidity. Five packets and five tin cans of seeds from each storage condition were prepared at the end of each of the four storage periods, to allow for testing after from two weeks to six months at 30°C. In a few cases, seeds from packets and tin cans were also tested after nine and twelve months.

At each testing period, duplicates of 100 seeds each were germinated (i) on moist filter-paper in Petri dishes at 20°C., and (ii) in the greenhouse in a mixture of equal parts of sand, sod soil, and granulated peat-moss.

In Table XIII are shown the germination percentages obtained, both at 20°C. and in the greenhouse, from seeds packeted after original storage periods of 3, 6, 12, and 24 months at 5° and 25°C., and then held for additional periods of 2 weeks to 6 months at 30°C. and 76 per cent relative humidity.

TABLE XIII

GERMINATION OF STORED ONION SEEDS AFTER PACKETING AND FURTHER STORAGE

Germination condition	Original storage condition				% Germination*	% Germination after further storage in packets at 30°C., 76% relative humidity									
	Time in months	Temp. in °C.	Open or sealed	% Moisture		Paper packets (months)					Tin cans (months)				
						0·5	1	2	3	6	0·5	1	2	3	6
			O		97	—	—	62	36	1	—	—	95	91	97
		25	S	10·6	96	—	—	41	14	0	—	—	94	91	97
			S	3·5	98	—	—	63	36	1	—	—	96	95	97
	3														
			O		97	—	—	64	29	0	—	—	96	95	97
		5	S	10·6	97	—	—	67	38	2	—	—	92	95	91
			S	3·5	98	—	—	62	44	1	—	—	94	95	97
			O		97	94	93	—	33	0	96	98	—	95	97
		25	S	10·6	91	92	77	—	3	0	93	94	—	93	95
			S	3·5	97	90	89	—	45	1	95	94	—	96	97
	6														
Moist filter-paper at 20°C.			O		94	93	92	—	40	1	99	95	—	97	93
		5	S	10·6	93	96	94	—	38	1	91	97	—	94	96
			S	3·5	91	95	95	—	45	1	97	93	—	98	96
			O		98	94	93	72	53	8	94	96	93	92	96
		25	S	10·6	93	86	79	6	2	1	90	94	75	77	83
			S	3·5	99	92	94	82	63	24	96	96	92	94	93
	12														
			O		95	98	93	78	57	16	97	98	96	94	92
		5	S	10·6	96	94	96	86	55	25	97	96	94	93	95
			S	3·5	98	92	95	77	56	26	97	99	98	94	91

		25	O		92	87	88	14	0	0	89	93	90	84	90
			S	10·6	76	24	2	0	0	0	62	71	61	53	59
			S	3·5	91	84	82	32	2	0	90	90	89	84	94
	24														
		5	O		96	88	93	17	0	0	92	97	96	93	92
			S	10·6	95	95	95	26	1	0	92	94	97	94	96
			S	3·5	97	94	90	29	6	0	95	96	95	92	93

		25	O		88	—	—	33	7	0	—	—	83	89	78
			S	10·6	45	—	—	20	4	0	—	—	79	76	45
			S	3·5	56	—	—	38	10	0	—	—	84	84	60
	3														
		5	O		84	—	—	29	12	0	—	—	78	86	71
			S	10·6	71	—	—	31	8	0	—	—	84	85	73
			S	3·5	68	—	—	28	17	0	—	—	83	85	67
		25	O		91	85	79	—	2	0	93	83	—	77	80
			S	10·6	87	81	64	—	0	0	81	85	—	81	68
			S	3·5	85	77	81	—	5	0	84	84	—	77	74
	6														
Soil in green-house		5	O		83	87	83	—	2	0	89	88	—	87	80
			S	10·6	80	84	71	—	7	0	93	81	—	77	75
			S	3·5	84	87	84	—	5	0	72	85	—	80	83
		25	O		85	93	70	61	25	1	88	84	84	78	81
			S	10.6	75	57	36	2	0	0	61	62	59	52	52
			S	3·5	83	87	85	72	28	3	90	90	85	85	72
	12														
		5	O		77	82	78	64	25	3	87	90	85	85	79
			S	10·6	92	85	85	62	33	5	84	88	90	76	76
			S	3·5	77	86	80	76	26	9	90	91	88	78	82
		25	O		86	81	68	12	1	0	86	81	86	81	77
			S	10·6	44	7	0	0	0	0	40	47	37	28	17
			S	3·5	81	81	63	9	1	0	83	86	86	87	89
	24														
		5	O		89	79	64	7	0	0	89	83	91	81	77
			S	10·6	85	69	73	17	0	0	69	79	87	91	77
			S	3·5	94	86	80	32	0	0	89	85	89	86	84

* Per cent germination upon removal from original storage condition for packeting.
— Indicates that no test was made.

It will be seen that almost complete capacity to germinate on moist filter-paper at 20°C. was maintained under all the conditions of the original storage up to and including twelve months. This means that for storage periods of one year or less, temperatures of 5° and 25°C., and open and sealed containers, were equally effective in maintaining a high quality in the onion seeds as measured by their ability to germinate immediately upon removal from storage. Furthermore, for germination under these conditions, there appeared to be no advantage in the initial drying of the seeds from 10·6 to 3·5 per cent moisture.

An extension of the storage period to twenty-four months, however, proved deleterious to the onion seeds containing 10·6 per cent of moisture when the storage temperature was 25°C. This will be seen in the reduction of the germination at 20°C. to 76 per cent. With this one exception, the germination capacity of all of the seed-lots immediately upon removal from the original storage conditions was high, ranging from 91 to 99 per cent. That there were differences in these seeds as regards their ability to withstand further unfavourable storage will be shown below.

The advantage of open storage at 25°C. over sealed storage of seeds containing 10.6 per cent moisture is evident. The atmosphere in the 25°C. storage chamber was dry enough to permit a moisture content lower than 10·6 per cent for the seeds in open containers. It always has a worse effect on seeds to seal them with excess moisture than to leave them in open containers in atmospheres of low humidity.

Reference to the data secured by testing these same seeds after further storage at 30°C. and 76 per cent relative humidity, reveals the striking superiority of sealed tin cans over the regulation seed-packets (Table XIII). Six months after packeting in tin cans, a very high capacity to germinate on moist filter-paper at 20°C. was retained by all lots, regardless of the original storage conditions or time, except by the one with reduced germination at the time of packeting, and even that lot still gave 59 per cent germination. An extension of the time in tin cans at 30°C. to twelve months failed to lower appreciably the germination capacity of onion seeds which had received a previous storage of six months at 5° or 25°C. This is in striking contrast to the fate of the seeds in paper packets, all of which were entirely worthless in six months, and were already so reduced in germination vigour after three months that good seedling stands could no longer be obtained in soil.

Most rapid degeneration in paper seed-packets is seen for the seeds containing an original 10·6 per cent of moisture and from an original storage temperature of 25°C. in sealed containers, regardless of whether the original storage time was for 3, 6, 12, or 24 months. This was to be expected as these were the most unfavourable original storage conditions used (Table XIII). These seeds germinated 96, 91, and 93 per cent respectively after 3, 6, and 12 months original storage, and were indistinguishable in this regard from the seeds stored under all of the other conditions. The inferior quality showed up, however, in the decreased resistance of the seeds to further unfavourable conditions—in this case paper packets at 30°C. and 76 per cent relative humidity. This response was similar to that obtained previously (Barton, 1939, 1941).

Reducing the moisture content of onion seeds from 10·6 to 3·5 per cent before storing in sealed containers at 25°C., made possible the preservation of their germination vigour for somewhat longer periods upon removal to 30°C. and 76 per cent relative humidity. Some of these effects, together with a comparison of germination capacity in soil and on moist filter-paper, and of seed-envelopes and tin cans for packeting, are shown in Fig. 11.

Several facts are evident. One of these is the increased tolerance to subsequent unfavourable storage of seeds stored with a low moisture content; another is the efficacy of tin cans for satisfactory keeping after packeting; and still another is the poor performance of weak seeds in the soil as compared to moist filter-paper. Again, it should be emphasized that only those seeds which had had previous dry storage, and which had

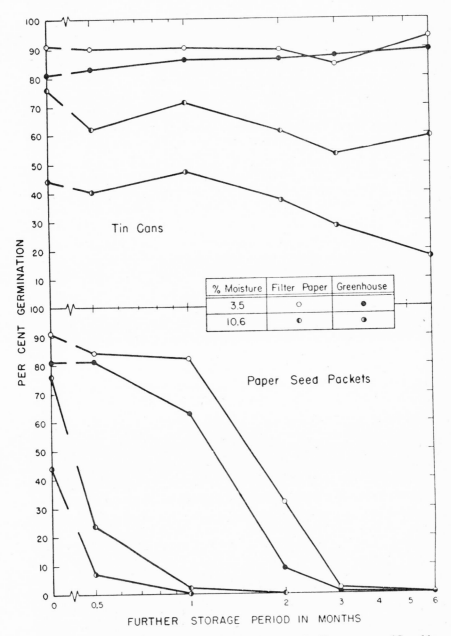

% Moisture	Filter Paper	Greenhouse
3.5	○	●
10.6	◐	◑

Fig. 11.—Graphs indicating percentage germination on moist filter-paper at 20°C. and in soil in the greenhouse, of onion seeds with moisture contents of 10·6 and 3·5 per cent held for twenty-four months at 25°C. and then placed in paper seed-packets or tin cans for a further storage period of one-half to six months at 30°C. and 76 per cent relative humidity.

therefore been able to retain more of their initial germination energy, were able to survive six months of packeting, even when air-tight packets were used. Low temperature will bring about similar effects when the moisture content of the seeds is unfavourable. Onion seeds with 10.6 per cent of moisture which had been stored for six or twenty-four months at 5°C. were much more tolerant of further storage in paper seed-packets than those previously stored at 25°C. for the same length of time. This difference was especially striking at the twenty-four-month period. Myers (1942) found that sealed lots of onion seeds stored in a refrigerator will germinate well on removal to 'natural' conditions for three months, while unsealed lots deteriorate rapidly.

A practical application of sealing in tin cans has been made by some seed companies among which is the Associated Seed Growers, Inc., of New Haven, Connecticut, who have described their 'canned seeds' in a monograph published in 1954 (Anon., 1954a). They point out that the moisture content of the seed requires no special adjustment if the seed stock is in a region of low average relative humidity. They made 'vigor ratings' of stored seeds and concluded (Anon., 1954a, p. 22) that 'it is feasible to preserve high initial viability and vigor of vegetable seeds by storage in moisture-proof sealed cans after the seed moisture content has been adjusted to a safe level which has been predetermined experimentally. Seed of onion, pepper, tomato, and certain other species is being Vigorpak processed, i.e. conditioned and hermetically sealed, and offered under the registered name of "Asgrow Vigorpak seeds".' Some of the apparatus they use is shown in Plate 6, and germination of cucumber seeds stored for two-and-a-half years in tin cans is pictured in Plate 7.

Air-tight tins were also found the most satisfactory containers for cucumber, lettuce, carrot, cabbage, and beet seeds in Queensland, Australia (F. B. Coleman & Peel, 1952).

From time to time, so-called moisture-proof plastic materials have appeared on the market. From these, water-resistant bags of various sizes have been made to protect such materials as foods and textiles, and parts of machinery, from the harmful action of excess moisture. Even though some moisture does penetrate the plastics, they have served the purpose of prolonging the usefulness of the encased materials in such places as the humid tropics. Barton (1949) tested several of these plastics to see whether they would protect seeds from excessive moisture. Eight different types of polyethylene and vinyl films allowed water to reach the seeds so that they could not be used for safe storage. Heat-sealed vinyl laminated aluminium foil, on the other hand, proved to be an effective moisture barrier and could be used safely to store onion seeds under conditions of high relative humidity. The importance of perfect sealing of the foil envelopes for safety in seed storage cannot be over-emphasized. If humid air is allowed to enter through an imperfect seal, the seeds acquire water readily and deterioration is hastened.

Photo. by courtesy Associated Seed Growers, Inc., New Haven, Connecticut, U.S.A.

PLATE 6

Apparatus which may be used for making a seed moisture-content test; tins and surrounding paper cartons from 60°F., 70°F., and 90°F. storage; electric hot-air oven for removing moisture from seeds for determination of seed moisture content. (From Anon, 1954a)

When one speaks of packets of seeds, the thought of small packets of vegetable or flower seeds usually comes to mind. However, packeting plays an important role in the life-span of other seeds, such as those of coniferous trees. Seeds of Douglas fir (*Pseudotsuga taxifolia* Britt.) and Western hemlock (*Tsuga heterophylla* Sarg.) present a problem to the seedsman and forester on at least three counts. First of all, frequent seed-crop failures make it difficult to maintain a constant supply of seeds for planting. Douglas fir trees, for example, produce a good crop every three to seven years, with light crops in between. The setting of good seeds in cones of Western hemlock is also likely to be sporadic. Another factor in the availability of seeds of these two species is their relatively short life-span under ordinary conditions of storage (Heit & Eliason, 1940; Anon., 1948). Still another problem, related to the second mentioned above, is that of retention of viability under shipping conditions. All of these difficulties are magnified by the tremendous importance of these trees, especially Douglas fir, for timber, and the consequent large demand for their seeds.

It has been shown in Chapter IV that sub-freezing storage is better than temperatures above freezing for maintaining the viability of many different kinds of seeds, including those of conifers. Furthermore, a recent work (Barton, 1954) has shown that $-18°C$. is better than $-11°$ or $-4°C$. for maintaining the viability of certain conifer seeds, including the two species considered here.

Some seed companies now use sub-freezing storage rooms, but the question of the behaviour of the seeds upon removal from the cold storage and packeting for shipment, and their ability to survive the shipment itself, have been matters of concern. Also, some customers have been reluctant to purchase so-called 'frozen' seeds.

We have already seen that the best condition for maintaining viability of onion seeds, as measured by testing immediately upon removal from storage, is the condition which permits the survival of the greatest number of seeds during periods of subsequent storage. Detailed results of such controlled tests on Douglas fir and Western hemlock seeds are now available (Barton, 1954*a*).

Seeds of Douglas fir and Western hemlock of the 1950 crop were received in March 1951. They had been kept at 0°F. from the previous December and were shipped to Yonkers, New York, by air express. Two temperatures, 5°C. and $-18°C$., were used for storage. Storage at 5°C. was in a large room with a fairly dry atmosphere for the first two years but with the atmosphere much more humid after that time. Sub-freezing, $-18°C$., was secured in a food freezer. Both temperatures were constant, varying only $±1°C$. Seeds with two different moisture contents were used.

When received, the Douglas fir seeds contained 5·8 per cent of moisture, calculated on the basis of the dry-weight of the seeds. Some samples were stored without any moisture adjustment. In other samples, the moisture content was increased to 13·6 per cent by spreading the seeds in a very humid room at 5°C., where they absorbed the desired amount of water in three days. They were then placed in a sealed container at −4°C. for one week in order to ensure uniform distribution of the moisture increase throughout all the seeds of the lot. Samples of the seeds with 5·8 per cent of moisture were placed in canvas bags for storage at 5°C. and −18°C. Other samples of the same lot, as well as of the lot with the moisture content increased to 13·6 per cent, were stored at the same temperatures after having been placed in tin cans with tight-fitting lids, further sealed with sealing wax. Each sample weighed approximately 90 gm. and consisted of about 8,400 seeds, one sample being intended for packeting tests after each original storage interval.

Moisture determinations made on seeds of Western hemlock upon receipt indicated that they contained 11 per cent. One lot was stored without any adjustment and another was dried to a moisture content of 7·7 per cent by spreading the seeds in the laboratory for four hours. Original storage in canvas bags and sealed containers at 5°C. and −18°C. was as described above for Douglas fir.

All moisture determinations were made by drying the seeds in a vacuum oven at 75°C. for forty-eight hours. Under sealed storage, the initial moisture contents were maintained for the entire test period. The seeds stored in canvas bags were subjected to moisture fluctuations corresponding to the humidity of the surrounding atmosphere. At the end of two years of storage, all lots in canvas bags at both −18°C. and 5°C. contained approximately 13 per cent of moisture. Subsequently, the atmosphere in the 5°C. room became much more humid, resulting in a moisture content of 22 to 26 per cent in the seeds by the end of the third year of storage. At −18°C., the seeds still contained about 13 per cent of moisture after three years.

Samples of both species were removed from the original storage conditions after 6, 12, 18, and 24 months and packeted in manila envelopes, foil envelopes, and tin cans for further storage at 5° and 30°C. Foil envelopes were made of vinyl-coated aluminium foil sealed by heat. Tin cans were sealed by tight-fitting lids and sealing wax.

Storage of the 5°C. packets was in a large room, the atmosphere of which was dry at the beginning of the test. When the atmosphere of this room became humid at the end of two years, all the 5°C. packets were moved to a small, dry chamber maintained at this temperature. Storage of packets at 30°C. was in a dry room with good air circulation, resulting in a low moisture content (about 4·5 per cent) of the seeds in manila envelopes.

Tests of the germination capacity of the seeds were made at the time of removal from the original storage conditions for packeting, and after 0·5, 1, 2, 3, 6, 12 and, in some cases, 18 and 24 months in the packets.

The results of these tests have demonstrated once again that sub-freezing temperatures are to be preferred to above-freezing ones for the maintenance of high quality conifer seeds. It has been shown further that the better the original storage condition, the greater will be the value of the seeds when they are removed from storage for planting or packeting or shipment and, ultimately, germination. This invalidates objections to the use of 'frozen' seed. Experimental results have shown that Douglas fir and Western hemlock seeds stored at −18°C. for as long as two years retain their full germination capacity and do not suffer, by virtue of their storage, upon transfer to a higher temperature. On the contrary, they are more resistant to subsequent deterioration than seeds which have been stored above the freezing-point, and which will have lost some of their germination vigour. For example, Douglas fir seeds packeted in manila envelopes held at 30°C. after six months of original open storage, deteriorated at the same rate whether the original storage temperature was −18° or 5°C. The same relationship held after twelve months of original storage, though the actual germination percentages were somewhat reduced. After twenty-four months of original storage, however, the great decrease in germination which took place during storage at 5°C. made the seeds worthless for packeting, whereas those stored originally at −18°C. were still of high quality. In other words, any condition which prevents deterioration of the seeds in storage, whether that condition be sub-freezing temperature, reduced moisture content, or both, permits an increased delay in planting. Some seeds held at a temperature as low as 5°C., but in high humidity, are worthless after a few months, whereas seeds of the same lots stored at sub-freezing temperatures for years on end may be as good as fresh seeds in every way.

Six months of original storage at −18°C. *plus* twelve months in packets, or twelve months of original storage at −18°C. *plus* six months in packets, had the same effect on subsequent germination of Douglas fir seeds when the packets were stored at 5°C. Furthermore, the conditions were as favourable for retention of viability as an original storage period of eighteen months at −18°C. When the packets were stored at 30°C., however, there was a greater retention of germination capacity after twelve months of original storage at −18°C. *plus* six months in packets, than there was after six months of original storage at −18°C. *plus* twelve months in packets. The survival of seeds upon removal from original storage and packeting, then, is directly related to the time they are exposed to an unfavourable temperature or humidity—whether that time be during the original storage period or after packeting for sale or shipment.

XII

METHODS OF TESTING FOR VIABILITY

ACTUAL germination of seeds in soil in the greenhouse or field, or in some suitable medium in the laboratory in controlled-temperature chambers, has been the tried and true method for determining germination capacity. However, this is not always as simple as it might appear. In the first place, there is apt to be a discrepancy between the ability of seeds to produce a root under controlled conditions on a medium such as filter-paper or a moist paper towel, or even in sand, and the actual stand of seedlings produced in the field (Goss, 1931; Kjaer, 1945; and others). Some of these relations have been worked out, so that the field performance can be forecast from the results of a laboratory test. In the second place, many seeds have special requirements for germination, and obviously will not produce seedlings unless these requirements are met. For non-dormant seeds, this may involve a special temperature, exposure to light, or mechanical or chemical treatment of the seed-coats to make them permeable. For dormant seeds, it may mean an after-ripening period either in dry storage, such as is characteristic of certain grains as well as of lettuce and some other seeds when freshly harvested, or in a moist medium at a low temperature, which is a requirement of the seeds of many trees and shrubs.

Not only may there be special requirements for germination, but these may change with storage. For example, low-temperature pre-treatment is reported to be more effective for old *Pinus palustris* seeds than for fresh ones (Barton, 1930), and the germination of old garden seeds is claimed to be benefited by soaking in ethylene chlorhydrin (Ruge, 1952). Many seeds having specific germination requirements at harvest time will germinate over a wide range of conditions after dry storage for a few months. For general discussions of these special requirements for germination in seeds, the reader is referred to five general articles (Barton, 1939*d*, 1953; Barton & Crocker, 1948; Crocker, 1948; Crocker & Barton, 1953). Some of these special requirements have been described in sections of the present book dealing with specific seeds.

Even when no special treatment is necessary to bring about germination, it is essential that there be a standard method for testing seeds of commercial importance, in order that they may be tested or shipped to any place in the world. This has led to the development of international rules for seed testing. These rules are formulated and approved by the

International Seed Testing Association and are revised as found necessary from time to time. The latest of these is to be found in the Proceedings of the International Seed Testing Association for 1956 (Anon., 1956). These rules were adopted by a meeting of the Association in Dublin on 29 May 1953, and were revised at the Paris meeting on 8 June 1956. The original rules came into force on 1 January 1954, for the Southern Hemisphere, and on 1 July 1954, for the Northern Hemisphere. The amendments became effective upon adoption. Hence the earlier International Rules for Seed Testing were cancelled.

Not only are there international rules, but individual countries also have seed acts governing the testing and sale of seeds within their territories. These methods as applied to the United States have been set forth in a Manual for Testing Agricultural and Vegetable Seeds (Anon., 1952a). Much research by many investigators has preceded the establishment of such rules, but this work need not be described in the present volume.

Special requirements for the germination of certain seeds have not been met by some workers testing viability after storage under various conditions. For this reason, the seeds have been reported non-viable when, in fact, they were capable of germination. This has applied especially to seeds with impermeable seed-coats or dormant embryos. Germination of elm seeds, for example, is inhibited by temperatures above 25°C. when they are mixed with moist granulated peat-moss, and very few seedlings are produced in soil from fully viable seeds which have not been pre-treated. Pre-soaking in water for twenty-four hours, as well as pre-treatment in moist granulated peat-moss at 5°C. for one month, resulted in greatly increased seedling stands in soil in the greenhouse (Barton, 1939a). Unless these facts are known, it is apt to be assumed that the seeds, which are presumably short-lived, are non-viable. The life-span of elm seeds was discussed in Chapter IX.

Germination tests of conifer seeds are also often inadequate and do not express their true viability (Mirov, 1936). Furthermore, it may happen that types of seed which germinate easily without pre-treatment when they are fresh or when retaining full viability in storage, may require pre-treatment for germination of deteriorating samples. This has been demonstrated for several seeds, among which are those of the longleaf pine, *Pinus palustris* Mill. (Barton, 1935a). From a given storage condition, these seeds retained only half of their original germination capacity as measured from plantings made directly in soil in the greenhouse, but if the same seeds were pre-treated for one or two months in a moist medium at 5°C. before planting in the greenhouse, four-fifths of the original germination percentage was obtained. Also, treatment with chemicals to prevent attacks by micro-organisms has been reported in many instances to increase the seedling stand from old seeds to a greater extent than from fresh seeds (*see*, for example, Wallen *et al.*, 1955).

QUICK VIABILITY TESTS

Because of the special treatments necessary, and the consequent delay, in finding the germination capacity of seed-lots of vital importance to the seed trade and to planters, an interest developed in ways and means of estimating the performance of the seeds without the necessity of germination tests. Perhaps the earliest attempt to get a rapid estimation of the germination capacity of seeds was through the 'cutting' test. The seeds were simply cut open and the number of empty ones recorded. This gives a fairly accurate measure of germination expectancy when the seeds are fresh or of high quality (Barton, 1930; Viado, 1938), but permits no evaluation of deterioration in the embryo.

A variation in the cutting method was reported by Dorph-Petersen (1925) and used by Trautwein & Wassermann (1929) and Weisenfeld (1929). Used for cereals and known as the 'half-grain' method, it involved cutting off one end of the seed so as not to injure the embryo. The remaining portions of the seeds were then placed on a moist substrate where the cotyledons turned green within a few days if the embryos were viable. Zachariew (1939) used this method as late as 1939 for seeds of *Pinus sylvestris* L. An extra refinement was made when the seeds were fully excised before testing.

EXCISED EMBRYO TECHNIQUE

Because of the continued dormancy of the embryo in dry storage, many tree and shrub seeds require a period of from one to several months in a moist medium at low temperature, where they after-ripen, before germination performance at higher temperatures can be ascertained. Such delays in the determination of viability may cause considerable inconvenience as well as increase the consumer's risk of buying seed stocks which have not been tested. A rapid means of determining the germination capacity of such seeds has, however, been developed in the excised embryo technique.

As early as 1906, Crocker emphasized the importance of the seed-coat in germinating weed seeds, and pointed out that removal of the coats of some weed seeds removed the germination barrier so that results could be obtained within a short time. Harrington & Hite (1923) removed the coats from dormant apple seeds and found that, if they were kept moist, the cotyledons became green and some of the radicles started to grow. Flemion (1936) and Tukey & Barrett (1936) were the first to apply this method to a rapid determination of the germination power of dormant seeds, and hence to use it as a measure of the viability of such seeds. Working independently, these authors reported on the use of the method for determining the germination capacity of peach seed. The rapid technique as used by Flemion for many types of dormant seeds (Flemion, 1936, 1938, 1941, 1948) compared favourably with germination tests which followed after-ripening at a low temperature.

The general method which Flemion used consisted of removing the hard outer coat, after which the seeds were soaked overnight in the laboratory, before removing the thin inner coats. The 'excised' embryos were then placed on a moist surface, usually filter-paper, in diffuse light in the laboratory, where the greening of the cotyledons and extension of the radicle indicated viability. It is not necessary to place excised embryos on a nutrient substrate, as recommended by Tukey & Barrett (1936), in order to test their viability. If the embryos have lost their viability they fail to turn green, and will rot within a few days. Plate 8, taken from Flemion (1938), pictures the appearance of excised embryos of different germination capacities. Flemion (1948) has summarized the work done with the excised embryo method and has listed the species that have been tested, together with literature references. This summary is given in Table XIV. The original literature citation numbers have been changed to authors and dates in the table here presented. In addition to those listed, Treccani (1951) used the Flemion method for testing fruit-tree seeds. A comparison of the excised embryo and the tetrazolium methods of testing for viability will be given in Chapter XIII.

TABLE XIV

LIST OF SPECIES, WITH LITERATURE REFERENCES, WHICH HAVE BEEN TESTED BY THE RAPID VIABILITY METHOD

Plant family	Species	Reference under literature cited
Amaranthaceae	*Gomphrena globosa*	Heit (1943)
Berberidaceae	*Berberis thunbergii*	Heit & Nelson (1941), Heit (1943)
Bignoniaceae	*Bignonia (Campsis) radicans*	Flemion (1941), Heit & Nelson (1941), Heit (1943)
Boraginaceae	*Cynoglossum amabile*	Heit (1943)
Caprifoliaceae	*Symphoricarpos racemosus*	Flemion (1941)
Celastraceae	*Celastrus scandens*	Heit & Nelson (1941), Heit (1943)
	Euonymus sp.	Heit (1943)
Compositae	*Coreopsis*	Heit (1947)
	Wyethia scabra	Flemion (1941)
Cornaceae	*Cornus* sp.	Heit (1943)
Cucurbitaceae	*Cucurbita, Cucumis*, and *Citrullus* spp.	Heit (1943)
	Echinocystis	Heit (1947a)
Elaeagnaceae	*Shepherdia argentea*	Flemion (1941)
Eucommiaceae	*Eucommia ulmoides*	Flemion (1941)
Euphorbiaceae	*Euphorbia marginata*	Heit (1943), Heit (1943a)
Hamamelidaceae	*Liquidambar styraciflua*	Anon. (1941c), Heit & Nelson (1941), Heit (1943)
	Hamamelis spp.	Flemion (1938), Heit & Nelson (1941), Heit (1943)
Leguminosae	*Cercis canadensis*	Flemion (1941)
Lythraceae	*Lagerstroemia indica*	Flemion (1941)
Magnoliaceae	*Magnolia acuminata*	Heit (1943)
Martyniaceae	*Martynia*	Heit & Nelson (1941), Heit (1943)

TABLE XIV—Contd.

Plant family	Species	Reference under literature cited
Oleaceae	*Fraxinus* spp.	Anon. (1941c), Heit & Nelson (1941), Heit (1943), G. P. Steinbauer (1944)
	Chionanthus spp.	Flemion (1938), Anon. (1941c), Flemion (1941), Heit (1943)
	Menodora spp.	Flemion (1941)
Pinaceae	*Abies concolor*	Heit (1943)
	Pinus spp. (13 species)	Flemion (1938), Baldwin & Flemion (1941), Flemion (1941), Heit (1943)
	Pseudotsuga sp. (Douglas fir)	Flemion (1938), Flemion (1941)
	Tsuga canadensis	Heit (1943)
Rosaceae	*Chaenomeles* (*Cydonia*) *japonica*	Anon. (1941c), Flemion (1941), Heit & Nelson (1941), Heit (1943)
	Crataegus spp.	Flemion (1938), Anon. (1941c), Flemion (1941), Heit & Nelson (1941), Heit (1943)
	Lyonothamnus floribundus	Flemion (1941)
	Malus spp. Apple	Flemion (1938), Heit (1943), Heit (1946)
	Prunus spp. (9 species)	
	Apricot	Heit (1946)
	Cherry	Flemion (1941), Heit (1943), Heit (1946)
	Peach	Flemion (1936), Heit & Nelson (1941), Heit (1943), Tukey (1944), Heit & Legnini (1945), Heit (1946)
	Plum	Flemion (1941), Heit & Nelson (1941), Heit (1946)
	Pyrus sp. Pear	Flemion (1938), Flemion (1941), Heit (1946)
	Rhodotypos kerrioides	Flemion (1938)
	Rosa spp.	Heit (1943)
	Sorbus aucuparia	Flemion (1938), Anon. (1941c), Flemion (1941), Heit (1943)
Sparganiaceae	*Sparganium eurycarpum*	G. P. Steinbauer (1944)
Umbelliere	*Trachymene* (*Didiscus*) *caerulea*	Heit (1943), Heit (1947a)
Urticaceae	*Cannabis sativa*	Flemion (1938)
	Ulmus americana	Barton (1939a)
Verbenaceae	*Callicarpa purpurea*	Flemion (1941)

A variation of this technique was used by Krstitch (1951–52) for embryos of *Pinus nigra*. They were placed on gelatin containing tannin, in which circumstances good embryos grow and turn red in twenty-four to forty-eight hours at 23° to 25°C. but dead or weak embryos have a brownish colour and are not turgid.

More recently Heit (1955) has described the embryo excision method as 'efficient, accurate and timely' for seed analysts. The seeds he tested included fifteen species of flowers and vegetables, twenty species of shrubs and vines, seven species of fruit trees, and forty-seven species of coniferous

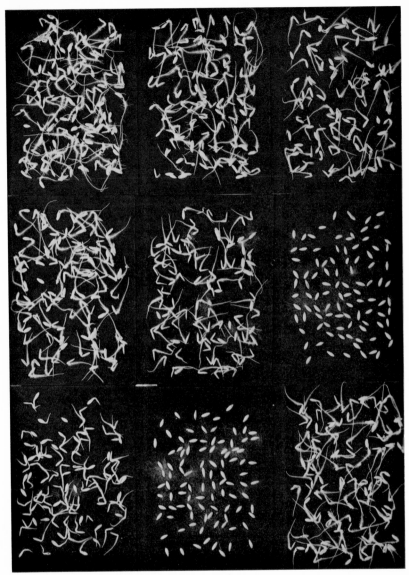

PLATE 7

Germinating cucumber seed-lots of different moisture levels from tin cans hermetically sealed and stored at three constant temperatures for two-and-a-half years. Pictures taken at time of first counts. From some lots, a few dead mouldy seeds have been removed. Germinations indicated in the table given below are totals of first and final counts, and are averages of four replicates of 100 seeds each. They range in columns from left to right in the illustration. Vigour ratings: E, excellent; G. good; F, fair; D, seed dead. Germination before storage was 98 per cent, and the seedling vigour was E. The results after two-and-a-half years follow:

Storage temp.	Moist. %	Germ. %	Vigour	Moist. %	Germ. %	Vigour	Moist. %	Germ. %	Vigour
60°F.	6.5	98	E	7·9	96	E	9·4	81	G
70°F.	6·5	94	E	8·0	86	G	9·3	0	D
90°F.	6·6	85	F	8·3	0	D	5·3	95	E

(From Anon., 1954*a*)

PLATE 8

Development of excised embryos on moist filter-paper in Petri dishes at room temperature: K, natural size; E, M, and N × 4; all others × 2. (Reading from left to right in each case.)

A and B. *Rhodotypos kerrioides*. A, Freshly excised embryo and five-day development on 10 November, 1937, of the 1932, 1934, 1935, 1936 and 1937 (2 embryos) crops, respectively. Note lack of growth in typical embryo from 1932 crop, limited development in 1933 crop, and increasing vigour in 1935, 1936 and 1937 crops. B, Embryos of 1932 and 1934 crops after ten days. The 1932 embryo is undergoing destruction by organisms and the 1934 embryo is still considerably behind the five-day development of embryos from later crops as shown in A. The 1934 embryos have a pale yellow-green colour, rarely exhibit any hypocotyl development, and deteriorate rapidly. Embryos which show this type of development are low in vigour, and only a low percentage of seedlings is obtained from them when the intact seeds are after-ripened at a low temperature.

C, Apple embryos. First, a freshly-excised embryo is shown and then (after six days on moist filter-paper) there are two non-viable ones followed by two viable embryos showing typical greening on cotyledons and new growth. Photographed 9 November, 1937.

D, *Crataegus crus-galli*. Ten-day development on 1 February, 1938. The first three illustrations are of embryos of the 1928, 1930, and 1933 crops, respectively. The embryo of the 1928 crop has undergone marked deterioration while the embryos of the 1930 and 1933 crops are also shown to be non-viable. At the same time the behaviour of two viable embryos is shown. At the extreme right in D is an embryo of a seed of the 1936 crop which had been in moist peat-moss at 5°C. for one year. Obviously the intact seed had not after-ripened during this time but had remained viable. The viable embryo of the 1935 crop (second from right) will, with time, also show similar enlargement of the cotyledon next to the moist filter-paper. It is readily seen that viable hawthorn embryos do not exhibit the extremely rapid development obtained with viable *Rhodotypos* seeds.

E, *Sorbus aucuparia* embryos. On the left is a photograph of a freshly-excised embryo, after which follows a dead embryo of the 1930 crop and a viable embryo of the 1937 crop after six days (18 November, 1937) on moist filter-paper.

F. Here are shown (18 November, 1937) embryos of Witch-hazel of the 1934, 1935, and 1936 crops, respectively, after five days. The 1934 embryos are dead and are undergoing deterioration, the 1935 embryos which are low in vitality exhibit very little activity—the cotyledons merely spread apart—while the viable 1936 embryos are growing. In G, H, and J are shown in each case two typical embryos of the 1934 and 1936 crops, the older embryo appearing on the left. The embryos in G have been soaked for twenty hours in a saturated solution of para-dinitrobenzene followed by one-half hour in dilute ammonia, and those in H have been treated for twenty hours in 1 per cent potassium tellurite. J illustrated the behaviour of embryos after five days on moist filter-paper. The chemical tests give some colour even in the 1934 crop, and reactions intermediate between those shown also occur, so that it is difficult in many cases to decide by chemical tests which embryos are alive. The results on moist filter-paper are much easier to interpret.

K. *Prunus americana* embryos after several weeks on moist filter-paper. All these embryos are viable and the various types of development are illustrated. The dead (not shown here) embryos deteriorate rapidly while the viable ones develop slowly.

L. Fringe-tree (*Chionanthus virginica*). The embryo on the left illustrates the deterioration of a non-viable embryo after four days, and the other photographs illustrate a viable embryo after four, six, and twenty days, respectively.

M. Douglas Fir. A freshly-excised embryo appears on the left, then viable and non-viable embryos after four days on moist filter-paper.

N. *Pinus rigida*. A freshly-excised embryo is seen on the left, and then a non-viable embryo and twe viable embryos exhibiting low and high vitality, respectively, after eight days.

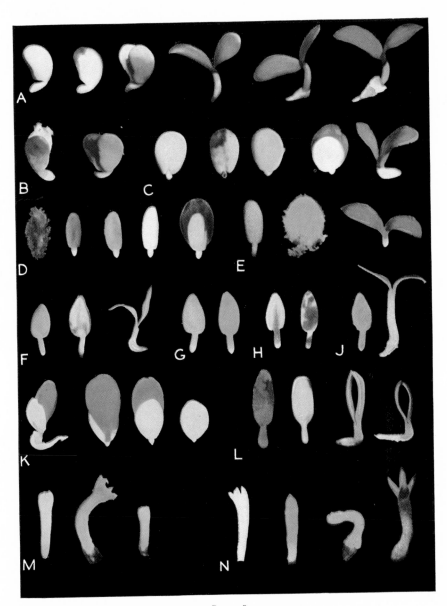

PLATE 8

and broad-leafed trees. Nord (1956) used 400 excised embryos, in lots of 100 each, for viability test of bitterbrush (*Purshia tridentata*) seeds.

Altogether, it appears that the excised embryo method is reliable and capable of more exact interpretation than staining tests. However, it apparently has not been included in any official, international seed testing rules.

PLASMOLYSIS

Plasmolysis, or the shrinking of the protoplasm from the cell-wall in the presence of solutions such as a saturated sugar solution or a 2-normal potassium nitrate solution, has been recommended for the determination of germination capacity (Doroshenko, 1937; Niethammer, 1942). But it has not been extensively used, owing perhaps to the skill required in preparation and in the microscopic determination of the germination values.

ENZYME ACTIVITY

With the realization of the need for rapid viability tests of seeds there came the idea of measuring their metabolic activity through some particular process such as respiration (Qvam, 1906) or the amount of heat given off when the seeds were placed under germination conditions (Darsie *et al.*, 1914). Biochemical assays also came in for their share of attention. One of the first, and the most widely tested, of these was enzyme activity. Much of the work in this field has been done on the relation of catalase activity to viability in seeds.

Crocker & Harrington (1, 18) were not able to use catalase activity as a measure of viability of seeds of *Amaranthus* and Johnson grass (*Holcus halepensis* L.). Nemeč & Duchoň (1922, 1923), however, found a close relation between germination capacity and catalase activity of oats from the crops of 1891 to 1912. Peas responded in a similar fashion. Niethammer (1931a) and Brodskis (1949) found a general high correlation between high catalase activity and high germination values. Leggatt (1929-30), in an extensive survey, found the catalase assay method promising. A statistical analysis of his data, made later (Leggatt, 1933), showed that viability, in the case of wheat grain, can be estimated fairly closely from a determination of total and thermostable catalase, but caution was advised on the routine use of such a method. These early reports are typical of the differences of opinion as to the value of the catalase-determination method of measuring germination capacity.

W. E. Davis (1925) pointed out that dry seeds, such as were used by previous workers, were not suitable for catalase measurements, as the enzyme does not necessarily disintegrate with the loss of viability but may remain unchanged for many years after the seeds become incapable of germinating. He postulated that catalase differences in non-viable and fresh seeds would be accentuated by placing the seeds under conditions which would initiate germination. He, therefore, soaked seeds in warm water at a temperature high enough to inactivate catalase, but not high enough to affect seed viability. Lettuce seeds soaked in water overnight at 32°C. showed reduced catalase activity in dead or partially deteriorated seeds, with no change in the amount of the enzyme in viable seeds. The catalase ratio, i.e. $\dfrac{\text{catalase activity of soaked seeds}}{\text{catalase activity of dry seeds}}$, rather than the actual catalase content, thus becomes the important consideration in determining viability. He noted, however, a tendency for the catalase of viable seeds to increase because of increased respiration following absorption of water. The amount of oxygen in the water used for soaking is also a factor in determining the amount of catalase activity. The conclusion from this work was that catalase activity could be an indication of viable as opposed to dead seeds in a sample, but could not be used to estimate different degrees of deterioration. It could be of value in the estimation of viability of dormant embryos in cases where dispatch is necessary. It could also be used to confirm results from other tests or to detect improper germination methods.

Takiguchi (1932) arrived at similar conclusions from a comparison of the catalase content of dried seeds and seeds soaked according to the Davis method. The 'catalase ratio' method of Davis was also used by Spaeth (1936) for tree seeds and by B. N. Singh et al. (1938) for seeds of Zea mays, Pisum arvense, P. sativum, Cajanus indicus, Triticum vulgare, and Cicer arietum, with the general result that ratios close to or above 1·0 indicate viability and ratios lower than 1·0 will indicate loss of germination capacity. However, as predicted by W. E. Davis (1925), decrease in catalase ratio was not always proportional to decrease in viability of these seeds at intermediate stages of seedling vigour.

The absolute amount of catalase in the resting seed of pine, in relation to the increasing amounts produced in seedlings within ten days of the commencement of germination, should be used to indicate the final percentage capable of germination, according to W. Schmidt (1929). Nuccorini (1930) agreed with this modification of the Davis method. Baldwin (1935) found that a few days' stimulation of conifer seeds in a Jacobsen germinator gave sufficient differences in catalase activity from the dry seeds to permit estimation of germination capacity. In this case, the ratio between the oxygen evolved by the germinating seeds and the oxygen evolved by the dry resting seeds, correlated well with actual germination tests. Quotients greater than unity indicated seeds of high quality.

It has been reported that catalase measurements are useful in the determination of damage to seeds by X-ray treatments (B. N. Singh, 1941) and by soaking in solutions of phenyl propionic acid (Nanda, 1950). Also, successful use of the method has been claimed by Petersohn (1926) to detect heat-killed white cabbage seeds as an adulterant in expensive red cabbage seeds. However, most of the people who have used this test have some reservations about its value in general use. Failure to get any correlation between germination and catalase measurements has usually resulted from the use of dry seeds alone (Gracanin, 1927; Grisch & Koblet, 1931; Knecht, 1931; Nazarova, 1937).

Other enzymes have also been shown to indicate viability, but mostly in connection with the staining of the living tissue. This will be discussed below. Direct measurement of oxidase content of cotton-seeds revealed that it controlled both the speed of germination and the water absorption capacity of these seeds (Nakatomi, 1936). McHargue (1920) reported a correlation between peroxidase activity and the viability of seeds of twenty species.

CHEMICAL TREATMENTS

External application of chemicals which bring about observable changes within the living tissues of the seeds was used at an early date. Perhaps the first was sulphuric acid applied on the cut surfaces of grain seeds, resulting in the formation of a deep rose-colour in living seeds within five minutes, while fifteen minutes were required for the appearance of the colour in seeds of poor quality (Dimitriewicz, 1876). Lesage (1922) used potassium hydroxide solutions for determining the germination capacity of seeds of *Lepidium sativum* within four hours.

DYES

Neljubow (1925) mentions previous methods of testing germination capacity without germination tests, including the measurement of carbon dioxide production by seeds which have imbibed water (Qvam, 1906), the method of Lesage (1922), and the catalase method; but he states that none of these was effective. Neljubow proposed that organic dyes be used. He found that indigo carmine and acid violet, as well as a mixture of neutral red and methylene blue, would stain dead embryos or dead parts of embryos, but would not stain living embryos or living parts of embryos. Viability as determined by these dyes agreed very well with the results of germination tests. Issatchenko (1931) confirmed these results, emphasizing that the embryos must be taken out of the fruit- or seed-coats before they are tested.

The indigo-carmine method has been found effective for seeds of Umbelliferae (Doroshenko, 1933), *Caragana arborescens*, *Pinus sylvestris*, and *Picea excelsa* (Shefer-Safonova *et al.*, 1934), *Aleurites* (Tskoidze, 1936), and *Pinus sylvestris*, *Picea abies*, *Pseudotsuga taxifolia*, and *Abies alba* (Hao, 1939). Grimm *et al.* (1928) described the method as very simple and used it in numerous trials. The seeds were soaked in water for one to three hours, until the outer envelope could be easily detached with forceps. After removal of the coats, the seeds were soaked in the 1 : 2,000 indigo-carmine solution for four hours at 26° to 28°C. They were then washed and their germinative power determined from the colour: seeds that became entirely coloured were incapable of germination, while those that remained uncoloured were capable of germination and yielded normal seedlings. Partly coloured seeds, including those coloured at one end or showing spots at both ends, had defective embryos and their survival after germination could be predicted by the extent of the staining.

Bismark brown is the best indicator for barley seed viability according to Kornfeld (1930) who found that, with it, all non-viable seeds became dark brownish-green while viable seeds showed scarcely a trace of the colour. Germination experiments with the grains which had become coloured gave no seedlings. Methylene blue used previously to indigo carmine (Turesson, 1922), and malachite green (Gadd, 1944), are other dyes which have been used successfully. A more complicated method of staining was used by Tanashev (1938), who combined cresol red with other chemicals such as phenolphthalein and xylenol blue. He noted that this improved the dye-indicator method. Niethammer (1931), using intact seeds of grasses and leguminous plants, obtained good tests for viability with methylene blue, neutral red, and orange G. The permeability of the seed-coats to the dye is a vital factor in the use of this method. Also, the permeability of some coats for specific dyes may change as the seed ages. Perhaps it was for this reason that tests with 0·1 per cent Congo red did not always show a good parallel between staining and germinability.

SELENIUM AND TELLURIUM SALTS

Sakata (1933) objected to the indigo-carmine method of Neljubow on the grounds that if the seeds were left in the solution too long, viable as well as dead seeds became coloured. He used tellurium salts instead. Although some other workers have used tellurium salts, the majority of them have used salts of selenium, which give essentially similar results. The method has been described by Eidmann (1938). Seeds from which the outer coating has been removed are soaked in water for twenty-four hours at room temperature and then further soaked in a 2 per cent solution of sodium biselenite ($NaHSeO_3$), adjusted to pH 4·5–5·0, for forty-eight hours. In contrast to the dyes which stain the dead tissue, the selenium compound stains the living tissue a permanent red by depositing elemental selenium. The embryonic tissue of dormant as well as non-dormant seeds is affected. Dead tissue remains uncoloured. The intensity of the coloration increases in proportion to the degree of vitality of the seed. This poses the question of judgement of the results and requires experience to be able to interpret the findings successfully. In general, selenium tests of good seeds give reliable results, but seeds of poor quality give variable results which do not agree closely with germination tests in the case of some conifer seeds (Hao, 1939a).

L. P. V. Johnson (1947), as a result of his study of the use of the sodium biselenite method on seeds of a wide range of agricultural plants, concluded that it was generally satisfactory provided the pre-soaking step was modified to include sixteen hours of pre-soaking and eight hours of aeration. He pointed out, however, that the relation between the biselenite reaction on one hand and seed germinability and seedling vitality on the other, must be established for each species before the results can be interpreted correctly. The method, suitably adjusted, could be used for evaluation of the seedling performance which could be expected from a given seed-lot. In the case of seeds which have been killed by *Fusarium* infection or by frost, the selenite method gives results which are too high, according to Gadd & Kjaer (1940). Lakon, who might be called the champion of the tetrazolium method to be described in the next chapter, also found that the selenium method was quicker than the germination method, gave the highest germination capacity, and was not affected by dormancy (Lakon, 1940).

The selenite method has also been found useful for tests of seeds of *Citrus* (Monselise, 1953), winter barley (Eggebrecht & Bethmann, 1939), and Karoo bush (Henrici, 1943).

Boiled, ground, mashed and finely cut seeds or embryos of *Pinus nigra*, *Picea abies*, and *Fagus sylvatica*, and boiled and unboiled expressed juices of these seeds, were not coloured by sodium biselenite, but the undamaged living seeds were coloured red (Wach, 1942). The living tissue is evidently required for the reaction. Plaut & Gabrielit-Gelmond (1949) presented a table of results of what they call the two most promising of the staining methods for testing viability, namely selenium or tellurium salts and 2,3,5-triphenyltetrazolium chloride. The table lists a considerable number of cereal, forage, and vegetable-crop as well as conifer seeds, which have been found to respond to one or both of these treatments. These authors believe that the results warrant further intensive study for routine testing procedures. Tellurites were considered more reliable than staining with methylene blue, catalase activity, or the presence of soluble sugars, for determining the vitality of seeds (Toniolo, 1949).

DINITROBENZOL

Gurewitsch (1935) developed a method of determining viability, which was based on the power of the living cell to reduce dinitrobenzol through respiration. The reduction products, nitrophenylhydroxylamine and nitranilin, are distributed differently in the tissues of the seed, as is indicated by the unequal intensity in the oxidation-reduction process. When ammonia is added to viable seeds, nitrophenylhydroxylamine gives a characteristic purple colour. The original tests were used successfully with wheat and rye grain, stained by solutions of *meta*-dinitrobenzol. Later, however, Ebes (1936) and Gurewitsch (1937) determined that the effects were due to the *ortho* and *para* isomers present as impurities in the *meta* compound, and that the latter alone was ineffective. Weise (1937) confirmed the work of Gurewitsch with wheat and rye grains and extended the method to beans and peas. It could not be used with asparagus or fruit-tree seeds.

RESAZURIN

The use of resazurin for viability tests has been developed at Rehovot, Israel, by Plaut and his co-workers (Plaut & Halfon, 1954; Plaut & Heller-Cohen, 1956; Plaut et al., 1957). Resazurin is blue in alkaline solution and red in an acid solution. It can be readily reduced to a red colour, followed by orange and finally white. A solution of 0.003 per cent resazurin was used for testing the seed of pea, bean, and cucumber (Plaut & Halfon, 1954). The embryos of viable seeds remained white, but those of non-viable seeds stained red or blue. In this concentration, the chemical does not kill the embryo, so that the same seeds could be tested for germination. It was found that all the seeds with white embryos germinated, while those in which the embryos stained blue did not. Tests on cereals were performed by removing the pericarp and immersing the grains in a 0·01 per cent resazurin solution for six hours (Plaut & Heller-Cohen, 1956). A blue stain of the coleoptile or of a major part of the germ (embryo) indicated that the seed was dead, whereas a red or pink stain indicated viable seed. This method has probably not been tested sufficiently for recommendation at the time of writing.

OTHER STAINING METHODS

Other biochemical methods for determining viability have included boiling in water with a drop of concentrated sodium hydroxide for from twenty to thirty minutes (Windisch—according to Hiltner, 1891); potassium permanganate reduction (Hibbard & Street, 1929); treatment with weak hydrochloric acid solution (Mazkov, 1936); treatment with a saturated alcoholic solution of guaiacol (Brücher, 1948); and treatment with solutions of phosphoric acid, benzidine, or hydrogen peroxide—in each case containing sodium chloride and followed by treatment with Skellysolve B (Gustafson, 1951).

FLUORESCENCE MEASUREMENTS

Kugler (1952) reported that fluorescent substances leach out of non-viable seeds of Raphanus sativus, Sinapis alba, and Lepidium sativum, but that no such substance can be detected in good seeds. Germ damage in wheat grain is positively correlated with the extent of browning as measured by the fluorescence of aqueous extracts (Sorger-Domenigg et al., 1955).

XIII

METHODS OF TESTING FOR VIABILITY (contd.)

TETRAZOLIUM SALTS

KUHN & Jerchel (1941) were the first to recognize the value of tetrazolium salts as excellent indicators of reduction in biological materials. Seeds, leaves, and roots of *Lepidium sativum*, fermenting yeast, and lactic acid bacteria, all became red after treatment with the colourless tetrazolium. This chemical has an advantage over dinitrobenzol, which produces an evanescent colour in the tissues and which is poisonous, and over sodium selenite which, though producing a permanent colour, is also poisonous. Further, tetrazolium salts are more stable than some of the dyes and work even under aerobic conditions. In addition to imparting a permanent colour, the tetrazolium salt has a low toxicity both for the plants and for the experimenter, and is one of the few organic compounds that is coloured in the reduced state. In the presence of viable tissue, the almost colourless tetrazolium salt is transformed by enzymes into the insoluble bright-red triphenyl formazan.

CEREALS AND GRAINS

Lakon, who had introduced the selenium method of testing seed viability, was quick to see the advantages of the new tetrazolium method, which he developed. A description of the Lakon method for small cereals and maize is to be found in his article on 'The topographical tetrazolium method for determining the germinating capacity of seeds' (Lakon, 1949). These rules for the use of tetrazolium were summarized in order to correct errors in the method as used by many people after a detailed description was published by Lakon in 1942 (Lakon, 1942, 1942a). The use of the method requires the preparation of the embryo of the seed, so that all those parts which are stained by the chemical, i.e. the living tissues, will be visible. Except in the case of oats, the pericarp of cereal grains is impermeable and opaque, and the embryo must therefore be removed with a sharp scalpel or needle before being sectioned longitudinally. In the case of maize seeds, a longitudinal section through the centre of the embryo is insufficient, because the lateral root primordia remain invisible. Pre-soaking of the seeds in water facilitates the removal of the embryo.

For small cereals and maize grain a 1 per cent solution of 2,3,5-triphenyltetrazolium chloride is used. This reagent is light-sensitive, and so should not be exposed to direct sunlight or to prolonged diffuse light, though solutions of it will remain unchanged for several months in the dark. The pH value of the solution should be between six and seven for the best results. Embryos extracted for testing must be free from starchy endosperm and pericarp fragments, and must be continuously submerged in the solution. The reaction is complete after from seven to eight hours in the dark at room temperature, following which the response can be accurately determined according to Lakon (1949). To be classed as viable, embryos must have the plumule and adjacent tissue-bearing root primordia stained. Figure 12 shows the aspect of staining for seeds of wheat, rye,

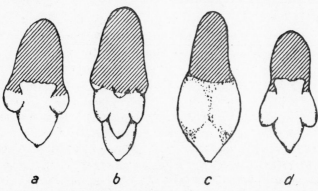

FIG. 12.—Excised embryos of cereals. (*a*) Wheat; (*b*) rye; (*c*) barley; (*d*) oats. Those parts of the embryo which, at the least, must show coloration after treatment with tetrazolium in order for the kernel to be germinable, are shown by hatch marks (from Lakon, 1949. Courtesy of Agricultural College, Stuttgart-Hohenheim, Germany).

barley, and oats. Maize grains differ in that germinable ones are stained either completely or at least in the region of the shoot, the stained parts including the initials of secondary radicles and the scutellum. Unlike the situation with other cereals, the staining of the scutellum of the maize seed is significant and at least half of its area must be stained to denote viability. Also, the tissue connecting the embryo and scutellum should be stained in its entirety. Lakon has pictured the minimum stained area required for a test of germinability of maize seeds in Fig. 13. Lakon warns against an extension of the test beyond twenty-four hours, for bacterial growth may appear and be stained by the tetrazolium and hence cloud the culture. In such cases the seed parts may appear to be stained improperly, being covered with minute red spots. Recognition of this condition is of prime importance, but the correct test of the seed tissue can still be determined by removing these spots by gentle abrasion of the surface of the tissue underlying them.

Many workers have been concerned with the staining of grain seeds, because of the necessity of early germination-appraisal after harvest and of the tendency shown by many of these forms to develop a dormancy during maturation. This dormancy is broken naturally by a period of dry storage, but requires from one to several months before germination can be obtained speedily under ordinary germination conditions. Cottrell (1947) tested cereal seeds by the tetrazolium method and was able to evaluate their germination capacity, either as a final or as a screening test, within twenty-four hours—as compared with ten or more days for a regular germination test. However, he found the method more exacting and time-consuming for the worker than ordinary germination testing.

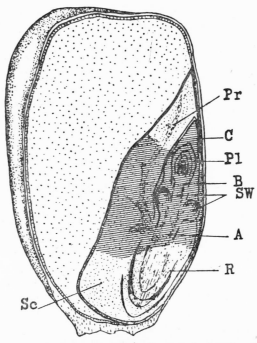

Fig. 13.—Median longitudinal section through a kernel of Maize. Sc, scutellum; Pr, procambial strand; C, coleoptile; Pl, plumule; B, base of plumule; A, transition from plumule to radicle; R, radicle; SW, initials of secondary radicles. That portion of the embryo which, at the least, must show coloration after treatment with tetrazolium in order for the kernel to be germinable, is shaded by hatch marks (from Lakon, 1949. Courtesy of Agricultural College, Stuttgart-Hohenheim, Germany).

The scope of the test is limited by the size of the seeds (Cottrell, 1948). New seeds of barley, oats, and wheat, incubated in a 1 per cent solution of 2,3,5-triphenyltetrazolium chloride at 45°C. for five hours, gave a good colour-reaction which appeared to be a quick, reliable index of germinability, but for old seeds of less than 60 per cent germination capacity, the test was less accurate, according to Shuel (1948). Favilli (1950) used 2,3,5-triphenyltetrazolium bromide in addition to the chloride to examine corn, wheat, rye, kidney bean, and lupine seeds, reporting an indication of viability which was 3 per cent greater than that secured in direct germination tests. He lists 119 items of literature on the subject.

There was an average difference, between sand germination and vigour as indicated by tetrazolium staining, of 7 per cent in tests made on wheat, barley, and rye seeds (Svenson, 1952) having germination capacities of from 80 to 100 per cent. Seeds showing intermediate colours described as pale pink, pink, milky red, or bluish, were interpreted as incapable of germination, as the tetrazolium indications were here usually higher than the germination tests. This emphasizes the importance, and difficulty, of interpretation in many cases. When the germination capacity of the seeds was less than 80 per cent with the sand method, the differences between the two methods was increased, resulting in discrepancies of from 30 to 40 per cent when the germinating capacities dropped to 30 or 40 per cent (Svenson, 1952). Toniolo & Beni (1954) also noted that the tetrazolium tests were good for seeds of high quality only. The method has been used on grains by many other people, among whom are Fink & Schweiger (1950), Bernal (1953–54), Ritvanen (1953), and Marras (1955).

Rice also has a period of dormancy following harvest, and the triphenyl-tetrazolium bromide tests have been made on these dormant seeds as well as on rice seeds which have lost their germination capacity (Niles, 1954, 1955). It is possible in this way to determine the eventual germinability of dormant seeds immediately after harvest, and so rice unsuitable for seed purposes can be diverted to other uses without the necessity of storage for periods of up to twelve weeks or even longer, and of periodic testing. The tetrazolium staining provided an accurate estimate of viability for the whole period extending from harvest time, through dormancy, and to the complete loss of viability (Niles, 1955).

Hyde (1949) has made a thorough study of the precision of the tetrazolium method in the estimation of germination capacity of seeds of Chewing's fescue. Using a special method of staining and of examining and classifying the embryos according to stain distribution, he obtained a high correlation with the percentage of viable seeds as indicated by a germination test. The standard error of estimated germination percentage from staining percentage was ± 5 per cent. When the tests were based on 300 seeds and the germination capacity was approximately 50 per cent, differences of more than 8·5 per cent were significant. Hyde concludes that the method is useful for seeds of Chewing's fescue and ryegrass (Hyde, 1949, 1952, 1955). In the case of ryegrass, there was close agreement between staining and germination tests for samples ranging in viability from 0 to 97 per cent, and the staining method proved more reliable when testing the seeds during the after-ripening period than at other times.

Lakon has been among the foremost proponents of the use of tetrazolium for testing seed viability, and has published many articles on the subject (e.g. Lakon, 1950, 1954; Lakon & Bulat, 1955). The method was received with enthusiasm by many workers confronted with the necessity of producing quick viability tests. The results of their experiments have produced various judgements of the value of the tests, from unqualified recommendation to complete rejection.

Tetrazolium staining was used successfully on the seeds of ten different kinds of grasses by Wharton (1955), and Davies & Winstanley (1957) found that it closely approximated the standard germination tests of seeds of Mexican grass (*Ixophorus unisetus*).

As was pointed out by Lakon, some of the failures may have been due to improper procedures, and variability in effectiveness was to be expected as the method was extended to many other types of seeds than grains. Also, as other seeds are used, special adaptations of the method must be made to fit the characteristics of the seed under study.

There has been a great deal of interest in the use of the tetrazolium staining method for determining freezing injury to maize seed, a subject of some economic importance. The method proved a reliable index to germination in the testing of four samples of artificially frozen, and one sample of naturally frozen, hybrid seed maize grain. These grains had been frozen at 48 per cent or less moisture and then dried to 12 per cent moisture before testing (Goodsell, 1948). The staining followed the use of 0·25 per cent tetrazolium solution to which was added 1 ml. of 10 per cent sodium hydroxide solution per 100 ml. The minimal damage of a hard freeze to viability of the seeds in a hybrid seed maize field could be determined within forty-eight hours if the seed ears were carefully shelled by hand and the seed dried rapidly to 12 per cent moisture with heated, circulated air. Isely *et al.* (1948) harvested maize seeds at various stages of maturity, subjected them to various sub-freezing temperatures, and tested them with tetrazolium chloride using standard germination in sand as a control. They found that tetrazolium gave higher readings in most cases, especially with high-moisture seed, than did standard germination in sand. Also, interpretation was difficult with frosted seeds and had to be made on the basis of differential staining rather than on the presence or absence of staining. These authors concluded, however, that the substitution of staining for germination tests is justifiable when immediate information is necessary.

Bennett & Loomis (1949) found the tetrazolium tests to be fairly accurate in assessing freezing injury to maize grain only if the germination was moderately high and the grain had been stored for some time after freezing. Necessity for storage may be related to the moisture content of the grain, which was 30 to 60 per cent in freshly frozen seeds, and which affects the staining test (*see* Goodsell, 1948). Bennett & Loomis (1949) point out, however, that it is possible to estimate the proportion of injured seeds immediately after freezing. This is done by observing the tint, texture, structural appearance, and other characteristics of the injured tissue—instead of the presence or absence of a pink stain—but this requires experience before a reasonable estimate of frost damage can be made.

A solution of 2,3,5-triphenyltetrazolium chloride can also be used to detect maize kernels killed by high temperatures, and to determine the percentage of dead seeds in a lot before purchase (Baird *et al.*, 1950). This was confirmed in 1957 (Schenk *et al.*, 1957), but at the same time it was further established that the presence of mould on the seeds gave a positive test. Differentiation between sound, mould-killed, and heat-killed maize seeds depended upon the texture of the embryo as well as on the colour test. Mould resulted in a friable texture, while embryos killed by heat were firm and rubbery. Tetrazolium chloride tests may give high values and be misleading in heat-damaged samples of barley grain, even though they predict the germinability of dormant barley seeds with accuracy (MacLeod, 1950). According to Verhey (1957), the tetrazolium method is not good for seeds injured by drying.

A weak response of maize, wheat, and cocklebur seeds to 2,3,5-triphenyltetrazolium chloride resulted from copper deficiency of the soil on which the plants were grown (J. C. Brown, 1954). It is to be expected that any environmental or nutritional condition which affects seed-set and development—as, for example, copper deficiency—will produce inferior seeds that are incapable of strong reduction of the tetrazolium in the tissues.

OTHER AGRICULTURAL SEEDS

The discussion so far has dealt with seeds of the monocotyledonous type, the structure of which appears suitable for staining tests. In general, less favourable results have been obtained with dicotyledonous seeds. This was found by Porter *et al.* (1947) for the seeds of leguminous plants. They attributed the trouble to the difficulty in sectioning the seeds, and to failures to detect possible seedling abnormalities—such as baldhead of beans. The seeds of both agricultural and vegetable crops were tested successfully by Mukherji (1956), using the tetrazolium method.

Wharton (1955a) compared the staining tests of 110 samples of seeds of various *Brassica* species with actual germination tests, finding the former satisfactory. Furthermore, the technique for these seeds is simple, permitting the examination of 100 treated seeds under the dissecting microscope in approximately twelve minutes. For each sample, four lots of 100 seeds each were placed in small dishes and covered with a 1 per cent solution of 2,3,5-triphenyltetrazolium bromide before being placed in the dark at 30°C. for forty-eight hours. The seed-coats were removed or slit for examination of the amount and distribution of the staining of the cotyledons and radicle. *Brassica* seeds often exhibit dormancy when freshly harvested, so requiring pre-chilling, light, or potassium nitrate treatment for germination at that time. This rapid staining test is useful for dormant samples or for non-dormant samples when a quick check on germination capacity is needed.

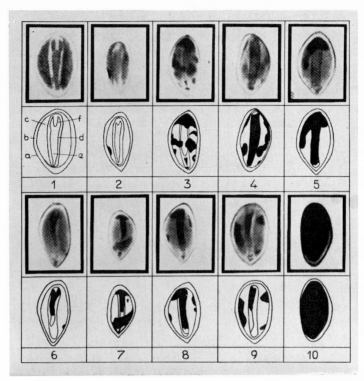

Photo. by courtesy Genetics Department, Forest Research Institute, Stockholm

PLATE 9

Various impregnation patterns of seeds of *Pinus sylvestris*. For each radiograph is drawn a sketch in which the impregnated areas are marked in black. 1. Non-impregnated seed: *a*, seed-coat; *b*, cavity between seed-coat and edosperm; *c*, endosperm; *d*, embryo cavity; *e*, embryo; *f*, cotyledons. 2 and 3 chiefly endosperm impregnations. 3. Note the small embryo impregnation near the micropyle. 4. Thin layer of impregnation in endosperm. 4 and 5. Spurious impregnation of embryo through influx of barium chloride into embryo cavity. 6. Impregnation of half the embryo. 7. Half of embryo now impregnated; the unimpregnated radicle is covered by an endosperm necrosis. 8 and 9. Various degrees of embryo and endosperm impregnation. 10. Wholly impregnated seed. (From Simak, 1957).

(*a*) Typical seedlings of Black Valentine Beans from favourable (*left*) and unfavourable (*right*) storage ten days after emergence. (From E. H. Toole, Toole & Borthwick, 1957.)

PLATE 10

(*b*) Typical plants of Black Valentine Beans from favourable (*left*) and unfavourable (*right*) storage at maturity. (From E. H. Toole, Toole & Borthwick, 1957.)

Photos by courtesy United States Department of Agriculture

Tetrazolium chloride also offers a rapid means of testing the viability of cotton-seeds, though any reaction other than complete staining of the embryo is indicative of loss of viability (Lambou, 1953).

TREE AND SHRUB SEEDS

It would be an advantage if dormant tree-seeds could be tested quickly by some less tedious method than the excised embryo technique. Nádvorník reported as early as 1946 that tetrazolium staining tests were feasible for the determination of viability of seeds of fruit trees and shrubs, good agreement being obtained between such tests and regular germination tests (Nádvorník, 1946).

An extensive comparison of the testing of both dormant and non-dormant seeds of varying viabilities of woody plants by the excised embryo and tetrazolium methods was described by Flemion & Poole (1948). They used 100 different lots of seeds representing fifty-eight species and seventeen families of plants. Detailed results of these experiments are shown in Tables XV, XVI, and XVII, taken from the article by Flemion & Poole (1948). The 'rapid viability test' referred to in these tables is the excised embryo technique. It will be noted that the tetrazolium tests were often characterized by different degrees of staining as well as by spot staining of all or some portions of the viable embryos. Dead embryos rarely became stained. The excised embryo tests also indicated two types of germination response, namely vigorous and sluggish. The data have been combined and rearranged to show general effects (Table XVIII). The authors explain this table as follows (Flemion & Poole, 1948, p. 251): 'The number of samples (column 1) with a particular type of staining (column 3) which showed various degrees of viability (column 2) by the rapid viability method are recorded. These are compared with the percentage of staining obtained. In 9 to 19 per cent of the 100 samples the viability test and the staining test agreed exactly (columns 4 and 5, lines 13, 9, 5, and 1, from the bottom). In a third to about a half of the tests the tetrazolium readings were from 1 to 30 per cent above or below the percentage viable obtained by the rapid viability test. In a considerable number of the tests the deviation between the two methods was greater than 30 per cent. For instance as in the case of embryos completely stained including all shades [Table XVIII] (line 9 from the bottom, columns 8 and 10), one-half of the 100 samples tested deviated more than 30 per cent above or below the rapid viability readings.' The correlation between the two tests is significant when all the data are combined in this way, but the variations in individual tetrazolium tests made some of them of doubtful value. The conclusion from these comprehensive tests is that more work would need to be done on an individual species before the staining test could be used as a measure of germination capacity.

TABLE XV

RESULTS OBTAINED WITH RAPID VIABILITY TEST AND STAINING WITH 2,3,5-TRIPHENYLTETRAZOLIUM CHLORIDE IN DETERMINING THE VIABILITY OF VARIOUS SEEDS

Family	Species	Rapid viability test			Tetrazolium test — Per cent of seeds giving colour										Deteriorated or empty seeds, not tested, %
		% vigor-ous	% slug-gish	Total % viable	Entire embryo stained			Most of embryo stained			Spot staining			No stain-ing	
					Deep	Med.	Light	Deep	Med.	Light	Deep	Med.	Light		
Aceraceae	*Acer tartaricum* L.	90	0	90		95					5				0
Anacardiaceae	*Rhus aromatica* Ait.	100	0	100	54			43			3				0
Berberidaceae	*Berberis thunbergii* DC.	77	6	83	49	21	27				36			1	3
Berberidaceae	*Berberis thunbergii* DC.	92	4	96		13	48				25			19	2
Berberidaceae	*Berberis thunbergii* DC.	30	4	34	37					19					0
Bignoniaceae	*Campsis radicans* Seem.	0	0	0										100	0
Bignoniaceae	*Catalpa speciosa* Warder	0	20	20										100	0
Caprifoliaceae	*Lonicera dioica* L.	78	0	78		10	74							9	7
Celastraceae	*Celastrus scandens* L.	2	0	2	11						4	24	24	31	6
Cornaceae	*Cornus* sp.	73	3	76	48	16					15			5	0
Hamamelidaceae	*Hamamelis virginiana* L.	0	34	34										100	0
Hamamelidaceae	*Hamamelis virginiana* L.	0	60	60										100	0
Hamamelidaceae	*Liquidambar styraciflua* L.	0	21	21										100	0
Leguminosae	*Cercis chinensis* L.	100	0	100	55	25		42						1	0
Leguminosae	*Robinia pseudo-acacia* L. (a)	54	0	54	32	25	2				16			18	0
Leguminosae	*Robinia pseudo-acacia* L. (b)	88	7	95	50	37		20				7		5	0
Leguminosae	*Robinia pseudo-acacia* L. (a)	65	6	71	9	34		42				8		12	0
Leguminosae	*Robinia pseudo-acacia* L. (b)	91	2	93	31			9							0
Malvaceae	*Abelmoschus esculentus* Moench (Okra)	77	8	85		83		7						1	0
Malvaceae	*Gossypium hirsutum* L. (Cotton)	81	14	95			95	7						1	4
Malvaceae	*Gossypium hirsutum* L. (Cotton)	0	0	0		1								94	6
Malvaceae	*Gossypium hirsutum* L. (Cotton)	63	17	80			64			12				16	8
Malvaceae	*Hibiscus syriacus* L.	96	2	98	100										0

Family	Species														
Oleaceae	Chionanthus virginica L.	56	22	78	24	42	28			3	2		4	1	0
	Chionanthus virginica L.	0	2	2					3	7	13	40		33	0
	Chionanthus virginica L.	54	9	63	70					9	21			2	0
	Fraxinus americana L.	63	28	91	21			19						3	0
	Ligustrum acuminatum Koehne	56	20	76	13	36		55			20		2	2	9
	Ligustrum ibota Sieb.	46	28	74	15	7	55	18					7	2	14
Rhamnaceae	Rhamnus japonica Maxim.	44	0	44	100										0
Rutaceae	Orixa japonica Thunb.	1	0	1							3			41	56
	Ptelea isophylla Greene	39	0	39	19	20	51							3	7
Taxaceae	Taxus sp.	77	0	77		21	60							19	0
Vitaceae	Parthenocissus quinquefolia Planch. (Virginia Creeper)	8	5	13								1	13	86	0

a = Permeable seeds; b = hard seeds.

TABLE XVI

RESULTS OBTAINED WITH GERMINATION TESTS, RAPID VIABILITY TEST, AND STAINING WITH 2,3,5-TRIPHENYLTETRAZOLIUM CHLORIDE, IN DETERMINING THE VIABILITY OF VARIOUS PINACEOUS SEEDS

Species	Germination test*	Rapid viability test			Tetrazolium test — Per cent of seeds giving colour										Deteriorated or empty seeds, not tested, %
		% vigorous	% sluggish	Total % viable	Entire embryo stained			Most of embryo stained			Spot staining			No staining	
					Deep	Med.	Light	Deep	Med.	Light	Deep	Med.	Light		
Abies balsamea Mill.	35	34	0	34	37			15	2				6	6	34
Larix europaea DC.	—	28	0	28	16	3	3	21				12		12	33
Picea excelsa Link	—	0	0	0										74	26
Picea excelsa Link	—	0	0	0										92	8
Picea excelsa Link	70	73	0	73	53	7		13			8			3	16
Picea glauca (P. canadensis B.S.P. var. glauca)	29	38	0	38	13	2		27		6				32	20
Picea pungens Engelm.	70	4	0	4	8	2								87	1
Picea pungens Engelm.	73	75	5	80	83		1	8		2	2			8	6
Picea pungens Engelm.	—	71	0	71	69		1							5	18
Pinus banksiana Lamb.	—	39	0	39	48	1	13	9		2	2	1		20	6
Pinus caribaea Morelet	64	11	1	12	46	2	32	15		2				38	11
Pinus caribaea Morelet	72	65	2	67	93			30						4	20
Pinus caribaea Morelet	—	86	0	86	88	3	1								6
Pinus echinata Mill.	—	92	0	92	28	3		21			1			37	8
Pinus echinata Mill.	—	40	2	42	51			16	4		4			4	10
Pinus palustris Mill.	—	72	2	74			3	19	4		67			4	21
Pinus strobus L.	—	18	20	38										92	7
Pinus sylvestris L.	—	42	1	43	34			12		18				24	8
Pinus sylvestris L.	—	64	3	67	57					1		6		11	24
Pinus sylvestris L.	—	76	1	77	70					4		1		5	13
Pinus taeda L.	46	45	2	47	32			22		3				4	42
Pinus taeda L.	—	16	3	19	13	2		10			3			24	48

Species											
Pinus taeda L.	—	—	8	2	10	3		7	3	55	32
Pinus thunbergii Parl.	—	—	27	2	29	30	4	12	9	39	6
Pinus virginiana Mill.	—	28	65	0	65	75	4	13		2	5
Pinus virginiana Mill.	—	—	23	2	25	7	1	68	9	9	7
Pseudotsuga taxifolia var. caesia Schwerin	—	—	0	0	0					92	8
Pseudotsuga taxifolia var. glauca Mayr.	—	—	0	0	0					84	16
Thuj occidentalis L.	—	—	0	0	0	6	2			91	9
Tsuga canadensis Carr.	—	—	1	0	1	6	2	2	7	81	2
Tsuga canadensis Carr.	—	—	0	0	27	40		6	3	4	45

* Seedling emergence in greenhouse soil.
— indicates that no test was made.

131

TABLE XVII

RESULTS OBTAINED WITH RAPID VIABILITY TEST AND STAINING WITH 2,3,5-TRIPHENYLTETRAZOLIUM CHLORIDE IN DETERMINING THE VIABILITY OF VARIOUS ROSACEOUS SEEDS

Species	% vigor-ous	% slug-gish	Total % viable	Entire embryo stained — Deep	Entire — Med.	Entire — Light	Most of embryo stained — Deep	Most — Med.	Most — Light	Spot staining — Deep	Spot — Med.	Spot — Light	No staining	Deteriorated or empty seeds, not tested, %
Chaenomeles japonica Lindl. (Japan-Quince)	30	14	44	65		13				20			2	0
Chaenomeles japonica Lindl. (Japan-Quince)	0	2	2										100	0
Cydonia oblonga Mill. (Quince)	0	31	31									9	90	1
Malus communis DC.	0	54	54	6		3	51		4	21	4		10	1
Malus floribunda Sieb.	0	33	33										100	0
Malus scheideckeri Spaeth	0	32	32										100	0
Prinsepia sinensis Oliv.	24	41	65								100			0
Prunus americana Marsh (Plum)	30	0	30	1	16	3		26	19	5		11	19	0
Prunus americana Marsh (Plum)	0	13	13									57	43	0
Prunus americana Marsh (Plum)	98	0	98			100								0
Prunus avium L. (Mazzard)	97	1	98		99		1							0
Prunus avium L. (Mazzard)	0	9	9		7				5				88	0
Prunus cerasifera Ehrh. (Myrobolan Plum)	5	6	11			89						3	8	0
Prunus cerasifera Ehrh. (Myrobolan Plum)	36	0	36		100									0
Prunus mahaleb L.	98	0	98				35	43	19	3				0
Prunus mahaleb L.	0	70	70								45	20	20	0
Prunus persica (L.) Stokes (Peach)	87	5	92	43	46		6	3					2	0
Prunus persica (L.) Stokes (Peach)	63	7	70		81		1						12	0
Prunus persica (L.) Stokes (Peach)	0	23	23	28								68	4	0
Prunus virginiana L. (Chokecherry)	0	0	0										100	0
Pyrus sp. (Kieffer Pear)	0	21	21	12	17	5	60	3		2			1	0
Rosa multiflora Thunb.	79	0	79	46	29	25								0
Sorbus aucuparia L. (European Mt.-ash)	97	1	98	98	19								4	0
Sorbus aucuparia L. (European Mt.-ash)	63	4	67	45						4				3
Sorbus aucuparia L. (European Mt.-ash)	0	62	62		6	41							53	0
Sorbus aucuparia L. (European Mt.-ash)	0	10	10			37							63	0

TABLE XVIII

COMPARISON OF RESULTS BETWEEN RAPID VIABILITY AND TETRAZOLIUM TESTS

No. of samples	Rapid viability test, % viable*	Staining standard	Tetrazolium tests Percentage difference between results of rapid viability and tetrazolium tests								Coefficient of correlation
			0%— (results identical)		1–30%		31–70%		> 70%		
			No.	%	No.	%	No.	%	No.	%	
45	61–100	Entire embryo stained	1	2	24	53	9	20	11	25	
29	21– 60	deep and medium	1	3	7	24	10	35	11	38	
26	0– 20	colour	9	34	2	8	0	0	15	58	0·7134
100			11	11	33	33	19	19	37	37	
45	61–100	Entire embryo	0	0	31	69	11	24	3	7	
29	21– 60	stained, all	0	0	7	24	8	28	14	48	
26	0– 20	shades	9	35	3	3	14	54	0	0	0·7949
100			9	9	41	41	33	33	17	17	
38	61–100†	Entire embryo	0	0	33	87	5	13	0	0	
22	21– 60†	stained, all	0	0	11	50	4	18	7	32	
40	0– 20†	shades	19	48	1	2	1	2	19	48	0·8630
100			19	19	45	45	10	10	26	26	
45	61–100	Entire embryo or	2	5	37	82	5	11	1	2	
29	21– 60	most of embryo	2	7	7	24	5	7	15	52	
26	0– 20	stained, all shades	10	38	1	4	1	4	14	54	0·8202
100			14	14	45	45	11	11	30	30	

* All of the embryos showing any development were listed as viable.
† Only embryos showing vigorous development were considered as viable.

Since 1948, the use of 2,3,5-triphenyltetrazolium chloride as a useful viability indicator has been reported for seeds of conifers (Parker, 1953; Yanagisawa & Asakawa, 1953), Beech (Parker, 1953; Evrard, 1955), and oaks (Parker, 1953). Hilf & Rohmeder (1955) give a detailed description of the penetration by diffusion of tetrazolium into the different cells of the embryo and endosperm tissues of forest seeds, and of the rapid reduction of the chemical in all tissues.

MECHANISM OF ACTION OF TETRAZOLIUM

Tetrazolium salt is an oxidation-reduction indicator, and the development of the non-diffusible red colour in a specific tissue is an indication of the presence of active respiratory processes in which hydrogen radicals are transferred to the tetrazolium chloride. The reduction of tetrazolium to the coloured formazan by living tissues is the result of enzyme action. C. O. Jensen et al. (1951) thought it probable that the reduction of triphenyltetrazolium chloride by viable maize embryos depended upon one or more dehydrogenase systems with favourable redox potentials. As we have seen above, the presence of active enzyme systems does not necessarily indicate viability, but the absence of dehydrogenase activity indicates inability to germinate. Roberts (1951) agrees with this explanation of the staining effect. The reduction of tetrazolium chloride in barley seeds is catalysed by a dehydrogenase, or dehydrogenases, with pH optima at 7·0 and 7·7 (MacLeod, 1952a). Although the loss of dehydrogenase activity in barley seeds paralleled the loss of viability, germination figures were apt to be lower than the values of active dehydrogenase indicated. Furthermore, enzyme activity persisted in heat-damaged barley seeds after complete loss of germinative ability. MacLeod concluded as a result of these tests that this staining technique is not a measure of life in the seeds but rather of the presence of active organic acid dehydrogenases.

F. G. Smith (1952) made a detailed study of the phenomenon and concluded that a positive tetrazolium reaction in the seed tissues may be due to the action of pyridine-nucleotide-linked dehydrogenases or possibly to aerobic dehydrogenases such as xanthine oxidase. A weak or negative reaction, Smith says, may indicate a deterioration or lack of the dehydrogenase or of diaphorase, which is thought to mediate (control) the reduction. An insufficient supply of substrates, or competitive action of aerobic hydrogen acceptors, may also result in a weak tetrazolium reaction. Under conditions where tetrazolium is used as a hydrogen acceptor, it may cause significant alterations in the respiratory metabolism (Throneberry & Smith, 1953).

In general, it can be said that staining with tetrazolium has an important role to play in rapid viability tests, especially in the cases of seeds of cereals, grasses, vegetables and other economic herbaceous plants; but the best method must be developed for each variety, and interpretation of results is much safer in the hands of experienced technicians than of others. A recent description of the general technique of tetrazolium testing has been given by Moore & Smith (1957). Fink & Schweiger (1950) give a table of results of tetrazolium tests by other workers. It has been generally conceded by workers in the field (see above citations; also Parker, 1955) that, although reduction of triphenyltetrazolium salts is not always a reliable indicator of the percentage germination to be expected, staining failure in cells normally reducing the chemical is a certain sign of deterioration. Whether it can be used as a general test for life in all tissues is still a question (Mattson et al., 1947).

That the tetrazolium method of testing viability has great potential importance is seen by the appointment of a Biochemical and Seedling Vigour Test Committee by the International Seed Testing Association. The 1953 report of this committee (Gadd, 1953) gave the results of experiments by seven different stations on seventeen samples of cereal seeds and one sample each of pea, bean, pine, spruce, and pear seeds. It was found that the test could be carried out with precision at the different stations and that, in general, similar results were obtained at these stations. The staining test did not reveal serious damage to rye grain or give information about low seeding value, but was a good index of sowing value of high-germinating seeds, and of one lot of medium germinating capacity. As a result of these co-operative tests, the committee concluded that they could not recommend the adoption of the method to replace the germination tests of cereal grains, though it could be used with success with other seeds. In fact, the use of the tetrazolium method for certain kinds of seeds was included in the Rules agreed upon by the International Seed Testing Association in 1953. These rules made it obligatory to include tetrazolium tests for viability of *Carpinus, Pinus cembra, Prunus, Rosa, Taxus,* and *Tilia,* and permitted their use for *Fraxinus* and *Juniperus.*

The 1957 report of the Biochemical Viability Test Committee (Germ, 1957) includes results of ten comparative tests on these seeds. The data from these tests and information from a questionnaire sent to seed-testing stations led to the following 'Synopsis and Suggestions' (Germ, 1957, pp. 320–1).

1. Biochemical tests are already made by a considerable number of seed-testing stations.
2. Amongst chemicals used for this purpose, tetrazolium-salts are taking first place.
3. The topography of the colouring is decisive for the judgment.
4. Uniform results can be achieved only if standardized methods are used. Special methods must be worked out for every kind of seed.
5. It is recommended to include the methodology worked out by Dr. Bulat in its rules and make it obligatory.
6. There exists no reason why the tetrazolium-method should be removed from the rules for those kinds of seeds for which it is now obligatory or permissible, as long as the method to be used for each kind is well defined in the rules.
7. It is recommended to permit the tetrazolium-test or make it obligatory in future only if the method for that particular kind has been exactly worked out.
8. The analysts who are carrying out the tetrazolium-test must be adequately trained.
9. It is recommended to draw attention to the difficulties of the tetrazolium-test either by the following sentence or some other sentence to the similar effect: 'The tetrazolium-test should be carried out exclusively by experienced and sufficiently trained analysts.'
10. It should be suggested to the seed-trade (FIS) not to finalize purchases based on tetrazolium-values, and to make reference to an arbitrary analysis if possible.
11. The latitudes of analyses as they are included in the rules for germination-tests can be used for the tetrazolium-tests only if these are carried out with 4×100 seeds. It is recommended not to depart from this rule in the tetrazolium-tests.
12. We have seen that it is possible to bring about germination of kinds of seeds which can otherwise be germinated only with great difficulty by certain types of preparation (*Tilia* and *Fraxinus*). Further efforts in this line should be made as they might lead to the discovery of some simpler and cheaper method than the tetrazolium-test of testing the vitality of difficult seeds.
13. The further activities of a similar Committee in this line should include de-tailed studies of the methodology applicable to those types of seeds which are

already included in the rules and new ones to be included, and the initiation of an enquiry amongst all interested members.

ELECTRICAL RESPONSES

Waller (1901) attempted to estimate the vitality of seeds by means of an electrical 'after-current' excited by living matter as the result of an 'exciting current'. Fresh, vigorous bean seeds gave a large response, while dead seeds or those of low quality gave little or no response. Similarly, good seeds killed by boiling produced no currents. In general the 'blaze' reaction of the seeds was parallel to the results of germination tests. Also, the correlation between the electrical reaction and the germinative capacity led to the conclusion that the magnitude of the reaction is some measure of vitality or vigour of germination. Fraser (1916), using the same method with seeds of barley, confirmed the results of Waller.

Fick & Hibbard (1925) adopted the method of measuring the relative outward diffusion of electrolytes (as indicated by conductivity readings) as an index of viability of timothy and red clover seeds. They considered this method superior to the measurement of resistance of the seeds themselves and to the comparison of the relative absorption and excretion of salts. This method was examined further (Hibbard & Miller, 1928) and found effective for distinguishing between seeds of high, low, and medium germinative capacity. However, these authors noted objections to the method. It is expensive and requires an operator with general knowledge of conductance measurements. It also presupposes that the electrolytic contents of all samples of one species are the same, which is doubtful, for the chemical compositions of seeds are known to vary according to the locality in which they are grown. The chemical composition of seeds is especially sensitive to nutrition, which varies greatly in different soils.

The 'prime' potential, or the reading obtained when the electrodes first touch the maize seed, is highly correlated with seed viability as measured by percentage germination and by yield (O. E. Nelson & Burr, 1946). The 'equilibrium' potential, which is reached after the galvanometer needle has fallen off rapidly for from 30 to 120 seconds, and which is retained for two to five minutes, is not related to seed viability, but rather to the genetic constitution of the plant. Seeds with superior growth-performance could be segregated regularly by this method, even though such growth was not manifested by the plant for from six to eight weeks. The electro-motive force of the seedling can be used to determine the inherent vigour of various maize hybrids and hence could be of great value in genetical work, as well as in determining the best prospects for survival in the nursery and garden (Anon., 1947).

X-RAY ANALYSIS

Perhaps the most recent of the rapid viability tests is the X-ray contrast method developed by Simak and his co-workers (Müller-Olsen *et al.*, 1956; Simak, 1957; Simak *et al.*, 1957). It has been used exclusively so far for conifer and other tree seeds (*see* above citations; also Evrard, 1957). The method is based on the penetration of the seed-coat and of the dead seed-tissue by barium chloride in hydrous solution, followed by a radiograph which reveals the necrotic, impregnated parts of the seeds in striking contrast to the living, unimpregnated parts. Various impregnation patterns of seeds of Scots pine (*Pinus sylvestris*) are shown in Plate 9 (Simak, 1957). Simak points out that photographic enlargements of radiographs are difficult to reproduce and hence sketches are included in the figure.

The X-ray method has an advantage over other quick viability methods which require the removal of the seed-coats, and, in many cases, some special preparation of the embryo for testing. Also, the X-ray contrast method offers a combined analysis of the structural and physiological condition of the seed. The former is ascertained by the relation of embryo and endosperm shown in the radiograph. As the seeds were not killed by the X-rays, it was possible to lay them out in the germinator in exactly the order shown in the radiograph, and thus consider the germination of every single seed on the basis of its anatomical structure and the degree of impregnation. Simak (1957) found a single seed of one sample of Scots pine which germinated in spite of its impregnated endosperm and cotyledons, thus showing the test to be not infallible, and he suggests that the X-ray test may be combined with the tetrazolium test. The two are complementary in their principles and thus a topographic classification for the X-ray method could be evolved. An impregnation ratio of one-fourth has been defined to mark the border between viable and non-viable seeds. Simak (1957) warns that barium chloride is not a universal impregnation agent for X-ray contrast analysis, and that seeds of different species may react differently to the same chemical substance.

XIV

PLANTS FROM OLD SEEDS

THE increasing practice among seedsmen of keeping surplus supplies of seeds for sale in subsequent years has resulted in part from the experimental determinations of the storage conditions effective for the maintenance of viability of the various types of seeds. For the most part, the measure of the keeping quality has been the power of the seeds to produce seedlings as determined by germination percentages obtained from soil or other plantings. However, the vigour of the plants produced from old seeds is a matter of increasing importance in view of this mounting practice of storing seeds for long periods.

There have been reports from time to time on the value of old seeds for planting, apart from their germination capacity. Bartlett (1859, p. 332) quoted from the *Irish Farmer's Gazette* that, 'The gardener knows that melon and cucumber seeds, if used of the last year's saving, produce plants too vigorous to produce much good fruit; whereas, those kept over for several years produce less rambling, but very fruitful plants'.

Ohga (1926; *see also* Plate 1) found that sprouts from Indian lotus seeds reputedly more than 1,000 years old were always longer than those from new seeds of this plant. A note (Anon., 1932) on trials to determine the relative values of pumpkin seeds of different ages, claimed that three-year-old seed gave the best results in one trial while, in another, four-year-old seeds yielded less quantity but better-quality pumpkin fruits than one-year-old seeds. It is not stated whether the yield was from a given number of seeds or from a given number of plants. Delayed germination, followed by decreased growth-rate and lack of a well-developed primary root, characterized old seeds of cucumber (Parkinson, 1948). Old seeds of lettuce also produce abnormal, very short seedlings with practically no roots, according to Drake (1948).

Beattie *et al.* (1932) determined the yield of peanuts from old seeds. They did not measure yield from the same number of plants in all cases, but rather the yield from the plants produced from a given number of seeds—regardless of the germination percentage. There were rather wide differences in the stand resulting from differences in percentage of germination, but these differences did not appreciably affect the yield. Beattie *et al.* attributed these results partly to soil heterogeneity and partly to the larger yields of plants adjacent to missing hills, but warned that these results should not be taken as an indication that the use of poor seeds is a profitable practice.

Kiesselbach (1937) used maize seeds of various ages and tested the yields obtained from planting them. He found that satisfactory results may be expected from well-matured, viable seed maize grain up to four years old—provided it has been kept free from external moisture and insect or rodent injury. He tested seeds from one to four years of age annually for five years. No significant differences were obtained in the comparative acre yield in bushels from seeds of various ages. Also, there were no material effects of seed age upon the vegetative development of the plant.

Dungan & Koehler (1944) found that a decrease in the yield from old seed maize grain was caused by a reduction in the number of seedlings established in the field, and to a lesser extent by decreased yield per plant. Three-year-old and one-year-old seeds of six different lots all produced perfect field stands, i.e. the number of plants was not affected, but the yield of the plants from the older seeds was 4·8 per cent lower than that from the one-year-old seeds. This reduction was extended to 7·8 per cent when the reduced stand was also allowed to influence the yield.

Leus (1930) studied the relation of farm-crop seeds to production and found that in wet-season culture in the Philippines the maize ear and mungo seed yield was greater from thirteen-month-old than from three-month-old seed, but in the yields of seed from cowpea and soya beans the reverse was true. In dry-season crops also the oldest maize seeds (thirty-two months old) resulted in the highest production of ears to the hill, but cowpea, mungo, and soya bean plants from seeds eight months old gave a higher production than those from seeds eighteen months old.

Crocioni (1934) conducted soil planting experiments with *Triticum*, *Brassica*, *Medicago*, and *Trifolium pratense*, comparing seeds of different ages and following the development of the plants in each case. Plants from old seeds were retarded in growth and less resistant to adverse conditions. Paralleling the age of the seed, he found that old wheat and *Brassica* plants were smaller and their yield was decreased. With leguminous plants, the decreased yield was less evident and became marked only in the cases of very old seeds.

Riddell & Gries (1956) suggested that variations in the growth-rate of spring wheats developing from seeds of different ages is related to the temperature obtaining during maturation rather than to the age or conditions of storage.

A tendency has been noted in cotton for older seed to yield less, but these differences were not statistically significant (Christidis, 1954). It was concluded that the effect of age seemed negligible for the first ten years of storage of these seeds. No consistent effect of age on earliness, lint length, or ginning out-turn was demonstrated, though extended storage beyond ten years may show some such effects. Age not only decreases the viability of the embryos, but reduces the growth of plants of tobacco from ten-year-old seeds (Čirkovskij, 1953).

The influence of the age of sugar-beet seeds on their seeding value was studied by Filutowicz & Bejnar (1954). They conducted hotbed, greenhouse, and field experiments on eleven varieties from seven successive crop years. The results showed that one- to four-year-old seeds did not differ in viability, but older seeds (five to seven years old) declined in germination capability. Associated with this decline was a later appearance of seedlings in the field; but there was no reduction in the value of beets produced from such seed.

BEANS

Rodrigo (1939), working with mungo (*Phaseolus aureus* Roxb.), used eleven- to thirteen-year-old seeds which had been stored in bottles sealed with paraffin wax at room temperature. As controls he used one-month-old seeds taken from plantings of the old seeds made previously. At maturity the plants were harvested and data were taken of the number of pods produced, the weight of the dry pods, the weight of the straw, and the weight of the seeds per plant. In garden plots he obtained a significant difference in plant yields of new and old seeds in favour of the latter. Old seeds also gave a greater pod-yield. The effect on the yield of beans was similar to that on the yield of pods—except in one case where the pod- and bean-yield of new seeds was found to be slightly greater than that of old seeds. He does not give the germination percentages obtained for the various seeds used. He offers two possible explanations: (*a*) the vigour of the seed expressed in yielding power is correlated with its inherent longevity; or (*b*) the seed requires some 'seasoning' or 'curing' before it attains the peak of its vitality.

A recent report by E. H. Toole *et al.* (1957) also deals with *Phaseolus* (three varieties of snap beans), using seeds of the same age (forty-seven months) and noting the effect on growth and production of seed storage under favourable and unfavourable conditions. The favourable storage conditions were in a naturally dry warehouse at Filer, Idaho, and the unfavourable storage was controlled at 19°C. and 57 per cent relative humidity at Mercedes, Texas. The seeds were germinated and grown under controlled conditions of nutrition, temperature, and light, in randomized blocks in a growth room. Seeds from unfavourable storage conditions were planted two days earlier than those from the favourable storage conditions, because of the difference in rate of germination and vigour of growth. The plants were grown until the oldest pods were ready for harvest, when the entire plant was cut at ground-level and various measurements of growth and yield were made. That there were marked differences in the rate and vigour of growth is shown in Plate 10 (*a*) and (*b*). Plants from the better seed-storage conditions reached the flowering and fruiting stages earlier, and gave a significantly higher yield, than those from unfavourable conditions—in spite of the earlier planting of the latter. Thus there are differences in vigour of the plants grown from snap bean seeds which have been subjected to different storage conditions for the same length of time.

Various Other Plants

It has been pointed out that six-year-old carrot, eggplant (aubergine), onion, tomato, and lettuce seeds (Barton, 1939*a*), and ten-year-old *Delphinium* seeds (Barton, 1939*b*), all produced normal plants, but no quantitative data were noted.

A carefully controlled experiment was undertaken in 1946 (Barton & Garman, 1946) to determine whether the age and storage condition of the seeds of China aster (*Callistephus chinensis* Nees), verbena (*Verbena teucrioides* Gill. & Hook.), pepper (*Capsicum frutescens* L.), tomato (*Lycopersicum esculentum* Mill.), and lettuce (*Lactuca sativa* L.), would have an effect on the yield of plants grown from such seeds. In order to have fresh seeds of the same genetic strain to compare with those which had been stored for longer periods, plants of the five species concerned were grown in 1944 from seeds of the original lots which had been stored several years before. From these plants seeds were collected, cleaned, dried, and stored in the laboratory until April 1945. All of the fresh seeds, excepting those of lettuce, were produced in the greenhouse, pollination being controlled. The lettuce seeds were produced in the field but no other lettuce flowered in the vicinity; hence the strain remained pure. Very fresh seeds of pepper and tomato were secured for planting in April 1945, from plants (also produced from old seeds of the original lots) which had been grown in the greenhouse during the winter of 1944–5.

The effect of various storage conditions on the germination capacity of aster and verbena seeds was reported in Chapter VI (Barton, 1939*b*). Those seeds with approximately 4 per cent moisture content that had been stored in tin cans at $-5\,^{\circ}$C. were used. Plants of both of these kinds were grown in the field in 1945 from seeds as follows: lot A, seeds produced in the greenhouse during the summer of 1944, and lot B, seeds from the original stored lots. Thus the seeds used to produce plants in lot A were approximately seven months old while those used for lot B were approximately nine-and-a-half years old.

The germination capacity of the stored seeds of pepper, tomato, and lettuce has also been described (*see* Chapter VI; Barton, 1935, 1939).

Pepper seeds which had been mixed with calcium oxide to remove one-half of the original moisture and then sealed in tin cans and stored at $-5\,^{\circ}$C. were selected for these tests. To produce plants for field performance tests in 1945, fresh and old seeds were used as follows: lot A, seeds produced in the greenhouse in the spring of 1945 from plants grown from the seeds stored in 1932, as described above; these seeds were harvested not more than two weeks before sowing for the field test; lot B, seeds produced in the greenhouse in the summer of 1944 from plants grown from the seeds stored in 1932 as described above; these seeds were about eight months old when they were planted for the field trials; lot C, seeds from the original lot described above, and stored in 1932. Thus it is seen that pepper plants grown in the field in 1945 represented those from very fresh seeds (lot A), those from eight-month-old seeds (lot B), and those from thirteen-year-old seeds (lot C), all being from the identical genetic strain.

Tomato seeds were similar in treatment to the pepper seed except that two original storage conditions were used instead of one. This was to permit comparison of the field performance of plants from old seeds which had been stored under favourable and unfavourable conditions, as well as of fresh and old seeds. Seeds of lot C had been stored for thirteen years in open containers in the laboratory. Lots A and B were produced from C in the same manner as described for the corresponding pepper seeds. Seeds of lot F had been mixed with calcium oxide to remove one-half of the moisture and then sealed in tin cans and stored at $-5\,°C$. for thirteen years. Lots D and E were produced from F in the same manner as lots A and B were produced from C.

In the case of lettuce, as for tomato, seeds from two original storage conditions served as bases. Both lots had been mixed with calcium oxide to remove one-half of the moisture, but lot B was stored in the laboratory while lot D was stored at $-5\,°C$. Lots A and C were produced in the summer of 1944 from plants which had been grown from lots B and D, respectively.

Seeds of all varieties were planted in soil in flats and placed in the greenhouse in April, 1945. As germination occurred and the seedlings grew, they were transplanted to individual pots from which they were later set out in the field.

The flowers produced by aster and verbena plants were taken as a measure of the effect of age and storage conditions of the seeds from which they were produced. The weights of fruits produced by pepper and tomato plants were recorded, and the lettuce heads were weighed at the time when a majority of them had reached the best stage for marketing.

ASTER

Plants from seed-lots A (seven months old) and B (nine-and-a-half years old) were set out in the field on 29 May 1945. The planting consisted of eight rows with twenty-two plants in each, arranged as follows:

ABBAABB————————→B

BAABBAA————————→A, etc.

The flower-heads appearing on the plants were cut from one to three times per week, and a record was kept of the number obtained from each plant. The first flowers appeared on 18 July. The experiment was terminated on 27 August, at which time most of the plants were dead or dying. Plants from lot A produced 3,461 flower-heads and those from lot B produced 3,378 flower-heads during the season (Table XIX). These totals indicate no significant differences in productivity of plants from fresh or old seeds. However, if earliness in flowering is considered, the plants from the old seeds (lot B) were ahead of those from fresh seeds (lot A) for the first two or possibly three weeks. This shows not only in the totals, but also in the number of flower-heads cut from each row, though the yield from the rows varied in lot A from 327 to 546 flower-heads and in lot B from 268 to 543 flower-heads for the season. There was considerable variation in the flowering of the plants in different parts of the field, but lots A and B were equally affected by good or poor growth conditions.

TABLE XIX

CUMULATIVE WEEKLY TOTALS OF ASTER FLOWER-HEADS CUT DURING THE GROWING-SEASON OF 1945

Row	Lot A from seeds 7 months old, after 1-7 weeks							Row	Lot B from seeds 9.5 years old, after 1-7 weeks						
	1	2	3	4	5	6	7		1	2	3	4	5	6	7
1	1	23	148	331	475	533	546		8	84	239	337	382	432	441
2	0	22	105	209	319	366	384		2	30	114	199	261	295	311
3	1	22	123	252	371	442	473		14	66	205	344	421	451	461
4	1	27	139	212	346	426	432		5	58	194	344	456	525	543
5	3	37	148	240	330	378	384		6	65	205	335	398	436	449
6	2	29	176	323	428	483	498		4	58	173	264	352	414	424
7	0	15	128	239	324	402	417		5	53	162	210	236	262	268
8	0	8	91	173	267	314	327		3	55	205	365	441	475	481
Totals	8	183	1,058	1,979	2,860	3,344	3,461		47	469	1,497	2,398	2,947	3,290	3,378

VERBENA

Because of the growth habits of these plants and the difficulty of securing exact data on their productivity, eighty-eight plants each of lots A and B were grown side by side. No observable differences in the two lots appeared during the growing-season of 1945.

PEPPER

Plants from the three lots of seeds described above were set out in the field on 29 May 1945. For the arrangement of plants in the field a standard 3×3 form for a Latin square was used, with reshuffling of rows, columns, and treatments to form ten blocks of nine plants each, consisting of three plants from each age of seed. The yield, measured in weight of pepper fruits produced in each of the ten blocks, is shown in Table XX. All fruits except the final harvest were removed from the plants when they were just beginning to turn red, and were weighed forthwith. Lots A, B, and C were produced from fresh, eight-month-old, and thirteen-year-old seeds, respectively.

TABLE XX

TOTAL PEPPER FRUIT YIELDS IN OUNCES FOR SEED-LOTS AND BLOCKS
FOR THE ENTIRE SEASON

Lot	Age of seeds	Block										Totals
		1	2	3	4	5	6	7	8	9	10	
A	Fresh	63	63	112	105	73	40	118	56	125	74	829
B	8 mon.	81	79	52	53	63	59	95	77	85	109	753
C	13 yr.	34	93	60	108	59	39	91	34	87	125	730
Totals		178	235	224	266	195	138	304	167	297	308	2,312

It is immediately apparent that the block or position in the field influenced the yield. A total of only 138 oz. was produced in block 6, for example, while block 10 produced 308 oz. of fruit. Differences in yield of plants from the three seed-lots appeared very small and of doubtful significance. An analysis of variance test of these data confirmed the above findings.

The totalities of fruits remaining on the plants were harvested on 22 September 1945, the individual fruits being graded for size and photographed immediately after harvest. It will be seen from Plate 11 that the most large fruits were harvested from lot A, the plants from fresh seeds, on this date. Weights of these fruits were included in the totals of Table XX.

A measurement of the heights of the plants made on 18 July also failed to reveal any differences in the lots. A somewhat greater number of fruits were produced by lot A.

TOMATO

It will be recalled that six lots of tomato seeds were sown for the measurement of yield as follows: lot A, fresh seeds, and lot B, eight-month-old seeds—both produced from the original lot C which had been stored for thirteen years in open containers at room temperature when the field tests were begun; lot D, fresh seeds, and lot E, eight-month-old seeds—both produced from the original lot F which had been stored for thirteen years sealed with calcium oxide at −5°C. when the field tests were begun.

The seedlings were set out in the field on 2 May 1945, except in the cases of nine plants of lot C which were planted in the field on 12 June.

This was necessitated by the failure of the seeds of this lot to give the expected germination percentage and the consequent need for a second planting. These nine plants were those of blocks 2 and 3. It may be seen from their productivity as compared with plants of lot C in the other blocks (Table XXI), that the delay in planting had no significant effect. Seeds of lot C were slow in germinating owing to their germination capacity being decreased by improper storage. Consequently the plants were slower in development than those of the other lots. This may be seen in Plate 12, which illustrates a typical plant of each lot on 22 May when they were transferred from the greenhouse to the field.

TABLE XXI

TOTAL TOMATO FRUIT YIELDS IN OUNCES FOR SEED-LOTS AND BLOCKS
FOR THE ENTIRE SEASON

Lot	Age of seeds	Block						Totals
		1	2	3	4	5	6	
A	Fresh	544	568	798	708	543	658	3,819
B	8 mon.	643	397	734	738	594	599	3,705
C	13 yr.*	299	203	339	470	260	274	1,845
D	Fresh	592	650	697	662	519	635	3,755
E	8 mon.	645	451	814	599	367	713	3,589
F	13 yr.†	608	712	646	680	478	795	3,919
Totals		3,331	2,981	4,028	3,857	2,761	3,674	20,632

* Stored in open containers in the laboratory.
† Stored with reduced moisture content in sealed containers at −5°C.

Fruits were harvested when ripe and the number of fruits and their weights were taken for each plant. Pickings were made on twenty-four dates during the growing-season. On 25 September, when there was danger of frost, all of the green fruits remaining were harvested. Their weights are included in the figures for total yield.

An examination of the total yield of ripe fruit from each lot (Table XXI) shows the decidedly inferior production capacity of lot C. Also, certain parts of the field were better than others for tomato fruit production as indicated by the differences in yield of the blocks. Each block represents a random reshuffling of rows, columns, and treatments of a standard 6×6 form for Latin squares. The six replications of blocks were equal to the number of treatments—in this case different seed-lots used for plant production—and also to the number of rows. One tomato plant was placed in each cell of the Latin squares.

An analysis of variance was made, using the weights of tomato fruits obtained from each plant in the field. The difference in lots was highly significant, as was also the difference between Latin squares. A detailed analysis revealed the source of the difference in lots. The yield of lot C *versus* lot F showed the advantage of a good storage temperature, −5°C., combined with drying and sealing, over storage in the laboratory. Both of these seed-lots were from the same original source, and both had been stored for thirteen years before they were planted. Thus it is demonstrated once again that *the actual age of the seed is of much less importance than the environment in which it is held*. Furthermore it was demonstrated that the vigour of lot C, as measured by the ability to produce high-yielding plants, was decreased. But from the fruits which were formed, good seeds were obtained (lots A and B). There were no significant differences in the weight of fruits from plants of lots A and B, and hence fresh and eight-month-old seeds from the same source were equally good. Thirteen-year-old seeds of lot F, and fresh and eight-month-old seeds produced from lot F, were all equal in value for use in field plantings. Thus the total variation in lots is to be accounted for by the poor yield of lot C.

LETTUCE

Lots B and D represent lettuce seeds stored for thirteen years in the laboratory and at −5°C., respectively. Lots A and C represent fresh seeds produced from lots B and D, respectively, and stored in the laboratory for about seven months before planting for field tests. Thus once again in this experiment we can see the effects of storage conditions of seeds of the same age upon the plants grown from them. Also, we can compare the performance of seeds of different ages from the same source.

A standard 4×4 Latin square form served as a basis for the field planting. Eight replications were made. Two lettuce plants were placed in each cell of the Latin square. The measure of lettuce plant yields differed from those reported for aster, verbena, pepper, and tomato. In the four latter cases, flowering and fruiting (i.e. the development of organs concerned with sexual reproduction) of the plants served for comparison. But as the commercial value of lettuce depends upon its vegetative growth, the sizes and weights of its leafy heads were measured. The seeds were of the variety Iceberg. All of the plants were harvested on the same date, 21 June 1945. The entire plot was screened to prevent damage by rodents.

A summary of the yield in weight by blocks is presented in Table XXII. A statistical analysis of these data showed a significant variation in lots, rows, and columns, and between Latin squares. The cause of the variation of the lots is strikingly demonstrated. In this case the temperature at which the seeds had been stored for thirteen years was without effect on the plants produced from them. The age of the seeds, i.e. whether they were seven months old or thirteen years old, had a marked effect and was in favour of the old seeds. It should be pointed out that the summer of 1945 was a poor growing-season at Yonkers, New York. There

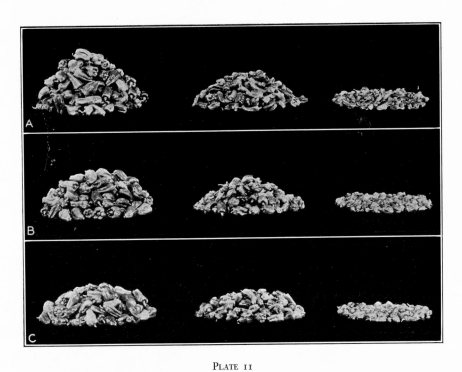

Complete harvest of fruits remaining on pepper plants 22 September, 1945. Each pile of fruits is 20 inches in diameter at the base. Top to bottom, lots A, B, and C, from plants produced from fresh, eight-month-old, and thirteen-year-old seeds, respectively.

was a great deal of rain and the temperature was low from 29 May to 7 June. Behaviour of seedlings in the field might vary with different weather conditions.

TABLE XXII

TOTAL WEIGHTS OF LETTUCE PLANTS, IN OUNCES, FOR SEED-LOTS AND BLOCKS

Lot	Age of seeds	Block								Totals
		1	2	3	4	5	6	7	8	
A	7 mon.	69	76	75	75	56	72	62	75	560
B	13 yr.*	76	88	93	73	66	86	77	82	641
C	7 mon.	76	87	61	75	58	69	79	69	574
D	13 yr.†	80	104	77	66	67	83	91	90	658
Totals		301	355	306	289	247	310	309	316	2,433

* Stored with reduced moisture content in sealed containers in the laboratory.
† Stored with reduced moisture content in sealed containers at $-5°C$.

All of the results for aster, verbena, pepper, tomato, and lettuce reported above were for the behaviour of plants which had been grown from the variously stored seeds and established in the field. In only one case, i.e. that of tomato seeds which had been stored in an open container in the laboratory for thirteen years, did old seeds germinate to form plants which were inferior in the field. This is not to say that there were no differences in the germination capacity of fresh and old seeds of the various plants. The particular lot of tomato seeds just mentioned, for example, gave only 6 per cent seedling production in soil in the greenhouse. This is to be compared with 85 to 97 per cent for all of the other lots—including the ones stored for thirteen years at a low temperature. The germination of the various lots of seeds was shown to have no direct relation to the field performance of plants grown from them. However, if the viability of the seeds is reduced too greatly, delay in germination followed by slow plant development will decrease their value for the production of seedlings of good quality, as was seen for tomato seed of lot C. Of course the actual number of plants produced from a batch of seeds, as well as the quality of those plants, is of great importance in commercial practice. If it is known, though, that good plants will be produced, the density of seeding can be adjusted to the known germination capacity. The data here presented would certainly justify the use of old seeds within rather wide age-limits.

As recently as 1951, Schwass (1951) indicated his belief that it is more economical in many cases to purchase certified seed of relatively low germination capacity, and that the resulting crop should be as good as from seeds giving high germination in the field, provided the sowing density is increased.

XV

CAUSES OF DETERIORATION

DIFFERENT explanations and theories of the cause of deterioration of seeds have been set forth. Regardless of the cause, ageing seeds result in abnormalities in the seedlings they produce. This is first evident in the delayed germination of old seeds and is often followed by the production of some chlorophyll deficiency or other abnormalities. Increases in the temperature and humidity of storage conditions for onion seeds increased the number of abnormal seedlings which failed to form sharply bent 'knees' and possessed blunt, undeveloped primary roots (B. E. Clark, 1948). This author stated that it was not clear from his results whether the abnormalities were purely anatomical or whether they were due to physiological deficiencies. It seems safe to say that both anatomical and physiological responses are involved, but that the mechanism of action has not been elucidated.

CHEMICAL COMPOSITION

Loss of viability has been thought by many people to result from a depletion of the food supply of the embryo. However, a cursory examination of just a few non-viable seeds convinces one that there is still plenty of reserve food left in the seed. A closely related view is that the reserves in the seed are changed chemically, so that they no longer furnish the nutritional requirements of the embryo. It has usually been assumed that any major chemical changes were in the embryo itself. However, Kikuchi (1954), using the method of transplanting embryos of wheat on to the endosperm, found that the older the endosperm was, the poorer became the growth of the transplanted embryo, though the embryo tended to age more rapidly than the endosperm in individual seeds.

Respiration is one measure of the metabolic activity taking place in the stored seeds. A low respiratory rate has almost certainly been attained by seeds that have a long life-span, whether in dry storage or in moist soil. On the other hand, increased respiration is apt to precede total loss of viability. The respiratory activity of wheat kernels was not affected by the length of the storage periods when tests were made after relatively short periods of six, eighteen, and thirty months (Leach, 1943). Increased moisture content and low quality of maize seed increases the respiration rate, according to Olafson et al. (1954). Oil-bearing seeds of high moisture content also have high respiratory rates (Qvam, 1906; Tashiro, 1913; Kretovich & Starodubtseva, 1956; and others). One of the problems in measuring respiration of seeds is the microbial population which may be present. For example, the respiration of soya bean seeds containing 18·5 per cent of moisture, measured after up to three weeks of storage, yielded curves similar in form to the microbiological population growth-curves (Milner & Geddes, 1945a). Obviously this difficulty is met most frequently in seeds of high moisture content. Gaseous exchange by seeds of very low moisture content, on the other hand, is much more difficult to measure because of the small quantities of gas involved.

The effect of storage of grain on its carbohydrate and protein content is of special importance because of the value of these constituents as food. Wheat grain stored for from three to fifteen years in a cool, dry, well-ventilated place in the laboratory was found to be suitable for starch production, though the recovery of starch decreased somewhat as the grain aged (MacMasters & Hilbert, 1944). Sugar loss during storage of different strains of sweet corn seed was about the same from year to year —indicating that the internal factors effecting the change from sugars to polysaccharides are inherited (Doty et al., 1945). There are small but definite differences in the in vitro digestibility of pure starches prepared from freshly harvested and well-stored seeds of rice, though there is no marked change in the amounts of the various carbohydrate constituents (Sreenivasan, 1939). Hemicelluloses and proteins decreased, and reducing sugars and sucrose increased, when expressed as percentage of dry-weight of blue lupine seeds stored at high relative humidities of 92, 86, and 80 per cent (Ward, 1951). This occurred during the first five months of storage and levelled off during the next five months. At lower relative humidities of 75, 65, and 54 per cent, the reverse was true, the available foods decreasing and the storage forms increasing. Depletion of reserve foods, i.e. hemicelluloses and proteins, was associated with loss in viability of the seeds. Acorns stored under normal atmospheric conditions show a minor loss of starch and a slight rise of sucrose, with little loss of viability (Serenkov & Kuznetsova, 1952). Stored in an atmosphere of carbon dioxide, the acorns show a loss of starch and a rise of sucrose and monosaccharides, the total carbohydrates remaining constant.

Crocker & Groves (1915) suggested that the degeneration of seeds in dry storage may be due to the gradual coagulation of the proteins of the embryo. Using Buglia's time-temperature formula for coagulation of proteins, they applied it to the deterioration of wheat seeds at two different moisture contents and various temperatures, and found that the calculated life-span of the seeds agreed well with the determined longevities for higher temperatures. Calculated longevities at lower temperatures, however, were much longer than the life-span of wheat seeds under ordinary conditions of storage, but they may well conform to the extended life of wheat grain under the best storage conditions. It remains for controlled storage of wheat grain to determine whether these calculations were reasonable.

Three different types of changes occur in the proteins of ground maize seeds and of whole shelled maize seeds during a storage period of two years (D. B. Jones et al., 1942): (1) a decrease in their solubility, (2) a partial breakdown of the proteins, indicated by a decrease in true protein content, and (3) a decrease in digestibility. Changes in the ground seeds were greater than those in the whole shelled seeds, and both were more affected at a storage temperature of 76°F. than were those stored at 30°F. Moreover, those in open storage were greater affected than those in sealed storage. Decrease in protein value and digestibility were more rapid during early stages of the storage. Essentially the same results were obtained for soya bean and wheat seeds (D. B. Jones & Gersdorff, 1939, 1941), and for peanut seeds (Moorjani & Bhatia, 1954). Since changes took place in the flour as well as in the intact grain, the proteins themselves, whether present in the embryo or the endosperm, are involved in the changes.

A study made by Fifield & Robertson (1952) on the milling, baking, and chemical properties of Marquis and Kanred wheat stored for from nineteen to twenty-seven years, revealed that storage had no consistent effect on the amount of protein. The best bread was made from the 1921 crop, probably because of its higher protein content.

Rice stored for from forty-six to eighty-four years in granaries in the northern part of Japan showed decreased amounts of fats in all cases and of sugars in some; but contents of protein, starch, crude fibre, and ash were no different from those of fresh seeds, though the power of germination was lost completely (Kondo & Okamura, 1934a).

The rather recent reports described above confirm the findings of Acton (1893), from a comparative analysis of wheat seeds twenty-eight years old and of new wheat seeds from the same field. In the old sample, insoluble compounds, such as proteins and starch, had undergone considerable changes to produce substances that were soluble in water. These changes suggest hydrolysis and enzyme action. The increase in the amount of dextrin and reducing sugars may have been caused by a slow action of diastatic enzymes, though no traces of these or of proteolytic enzymes were found in the old sample. The author believed it probable that such 'ferments' were originally present and produced the changes during the earlier period of storage, but were later destroyed by oxidation or by micro-organisms. This explanation seems to have been borne out by later studies which described changes taking place during the early part of the storage period.

ENZYMES

Because of their importance in metabolism, enzymes have been studied rather extensively in stored seeds, and have been claimed by some to be an index of the germination capacity and vigour of a seed-lot (*see* Chapter XII). Albo (1908) stated that the loss of germination capacity was directly related to enzyme activity. There is no change in enzyme behaviour in seeds of wheat or their milling products from harvest until the next spring, when dehydrogenase and peroxidase reached a minimum and moisture content was at a maximum (Kullen, 1941). High storage-temperature promoted, and low storage-temperature reduced, the changes in enzyme activity—except for catalase, which changed little in any case. Humidity was also a factor affecting enzyme activity.

Amylase activity in rice seeds is greatest at the 'milk stage', i.e. during the maturation of the grain on the plant, and decreases with advancing maturity and even more rapidly during the early period of seed storage (Sreenivasan, 1939; Rao *et al.*, 1954). The activity of amylase of acorns does not change when the seeds are stored in moist sand at 0°C., but storage in an atmosphere of nitrogen or of carbon dioxide increases the amylase activity (Evreinova & Erofeev, 1956). Amylase is important in the mobilization of starch, in respiration and growth of the seed.

The catalase content of old seeds, and the possibility of using this as a measure of viability, has been discussed (Chapter XII). However, it is unreliable as the results depend upon the seed and the special conditions under which the measurements were taken (Crocker & Harrington, 1918; Kondo & Okamura, 1931, 1934). W. C. Davis (1931) pointed out that catalase and oxidase were frequently used as criteria of metabolism and viability of seeds, and that, though the significance of catalase is a question, the oxidases are definitely associated with respiration and other physiological processes which may bring about deterioration. However, he could find no simple quantitative relation between phenolase activity and age or germination capacity; for though young wheat seeds of high germination capability showed high phenolase activity, and old seeds of low germination capability showed low phenolase activity, there was gradual loss of phenolase activity with age of oat grain—irrespective of viability—and this loss continued at the same rate after complete loss of viability. With cucumber seeds two, three, and four years old, both viability and phenolase activity decreased regularly with increasing age, whereas with barley grain from one to eleven years old, phenolase activity decreased but little, though germination dropped from 76 to 12 per cent.

Many old seeds were analysed for peroxidase by Brocq-Rousseu & Gain (1909). Using seeds of ninety-one genera, they found peroxidase in 1 more than 80 years old; 1 more than 76 years old; 11 more than 50 years old; 18 more than 25 years old; and 22 less than 25 years old. They stated that the maximum known persistence of peroxidase is 208 years.

VITAMINS

The vitamin content of old seeds is a subject of some interest. Thiamin content did not change in wheat grain during short periods of storage

(Pearce, 1943). However, extended storage of from five months to fifty-one years resulted in consistent losses of thiamin with increasing age (Bayfield & O'Donnell, 1945). Vitamin B decreased somewhat in rice grain with storage, but was still present in old seeds (Kondo & Okamura, 1932/33, 1932/33a). Increase in the time of storage and in storage temperature destroyed ascorbic acid in *Vigna sinensis* (Hoover, 1955).

AUXIN

The auxin content of maize seeds falls gradually with storage, but there is still a considerable amount present in twenty-six- to thirty-eight-year-old seeds (Juel, 1941). There is no indication that the disappearance of auxin leads to the death of the seeds.

ALKALOID

The alkaloid content of *Strychnos nux-vomica* seeds held in a gunny bag in the humid atmosphere of Dehra Dun, India, for sixteen years remained essentially the same (Puntambekar, 1947). It was concluded that the poor quality of such seeds for the manufacture of the drugs strychnine and brucine is due mainly to the initial low alkaloidal content of seeds from the same or closely allied species from different localities, and not to any deterioration caused by long storage.

ORGANIC ACIDS

There seems to be no doubt that an increase in free fatty-acid formation, especially in seeds with high oil contents, accompanies loss of viability, and also loss of commercial value in storage. This has been demonstrated for cotton-seeds and peanuts (F. R. Robertson & Campbell, 1933; Freyer, 1934; Hoffpauir *et al.*, 1947; Stansbury & Guthrie, 1947; Pons *et al.*, 1948; C. O. Jensen *et al.*, 1951; and others); for grains (Zeleny & Coleman, 1938; Slusanschi, 1939; Sorger-Domenigg *et al.*, 1955; and others); for sunflower (Lishkevich, 1953); for *Brassica* sp. (Täufel & Pohloudek-Fabini, 1954); for flax (Painter & Nesbitt, 1943); for sorghum and maize (Lindemann, 1953); for pine (Mirov, 1944); and for *Macadamia* (Chu *et al.*, 1953). The fat acidity test has been applied to several hundred samples of sound and damaged grain, with the result that it has been possible to establish fat acidity values for grain showing little or no deterioration (Baker *et al.*, 1957). It has been shown statistically that reduction in viability and citric acid content are parallel in cereal seeds (Täufel & Pohloudek-Fabini, 1955). Deterioration of seeds of *Medicago sativa*, *Trifolium pratense*, and *Lotus corniculatus* is due to an accumulation of lactic acid, according to Wyttenbach (1955). This lactic acid, probably produced as a result of intensified respiration, exercises a toxic effect on the embryonic tissues as soon as its concentration is increased. The dead grains contained at least fifty times as much lactic acid as fresh grains having a high germination capacity.

INHIBITING SUBSTANCES

Schwemmle (1940) offered some experimental evidence of the accumulation of inhibiting or toxic materials in seeds as they age. The seeds of *Oenothera berteriana* germinate more slowly and in lower percentages as they become older, up to eleven years. By running tests with fresh seeds which were treated with water extract or water-extracted mash of old seeds, he was able to show that the old seeds contained germination-inhibiting substances. The mash of the old seeds was especially effective in inhibiting the germination and later seedling growth of new seeds. *Oenothera odorata* seeds also lost germination capacity with age, but the development of the seedlings was much less inhibited by extracts of its old seeds than was the case with *O. berteriana*. Stubbe (1935, 1935*a*) has suggested that the metabolic changes of seeds in dry storage lead to the accumulation of inhibiting or toxic substances that transform the nuclear materials and lead to mutations.

Artificial mutations of seedlings of *Nicotiana tabacum* produced from normal seeds were induced by treating them with cold water extracts of ground, prematurely aged seeds. Oil and oily emulsions of the aged seeds had considerably less effect than the aqueous extract. Radiosensitivity of wheat seeds increases with increasing age, according to Nilan & Gunthardt (1956). The greater sensitivity of the chromosomes in aged seeds may be due to chemical compounds produced by the decomposition of food reserves.

MUTATIONS

There is a considerable amount of evidence that disarrangement of the nuclear mechanism is responsible for deterioration, or results directly from physiological changes which initiate the deterioration process. Shkvarnikov (1937) found an increased mutation rate in plants from six- to ten-year-old seeds of summer wheats as compared with fresh seeds. Mutations from old seeds showed chlorophyll deficiencies, thick heads, square heads, speltoid forms, dwarfs, and sterile plants. However, Shkvarnikov also found great variation in the mutation rate from seeds of the same age. It has been shown that seeds of *Datura* which were aged by storage in the laboratory for periods of up to ten years, have increased mutation rates for pollen-abortion mutations, and that these mutations are inherited (Cartledge & Blakeslee, 1934).

This work followed the discovery by Navashin (1933) that in root tips of plants grown from aged seeds of *Crepis*, chromosomal abnormalities were produced in large numbers. Shortly later Peto (1933) also found large numbers of chromosomal mutations in root tips of maize from aged seeds. The work on *Datura* has been extended by showing that high rates of visible mutations are also induced from old seeds (Blakeslee & Avery, 1934; Avery & Blaskeslee, 1936), and that the ageing of pollen grains for periods of up to thirteen days is even more effective for increasing the pollen-abortion mutations than is ageing of seeds (Cartledge *et al.*, 1935). Stubbe (1935) reported that ageing of seeds of *Antirrhinum* greatly increases the incidence of visible mutations.

These results raise the question of the causes of the increased mutation rates following ageing. That the increases are not due to age alone is indicated by the results of study of *Datura* plants grown from seeds that had been aged for twenty-two years in the soil (Cartledge & Blakeslee, 1935). These plants showed rates of pollen-abortion mutation that were only slightly higher than the rates found for their controls. Investigation of the effects on the mutation rate of varying the environmental factors to which the seeds might be subjected was thus suggested. Peto (1933) has shown for barley, and Navashin & Shkvarnikov (1933) and Shkvarnikov & Navashin (1935) for *Crepis*, that heating seeds is more effective than ageing for the production of chromosomal mutations in the root tips. Although many other accounts of experiments on heating seeds may be found in the literature, they seem to be connected with the control of fungal or bacterial parasites, or with the effects of the treatment upon the germinability of the seeds—or, in a few cases, with the abnormal growth of the seedlings developed from heated seeds.

When the moisture content is very low, seeds of many sorts will endure surprisingly high temperatures. Harrington & Crocker (1918) germinated seeds of Johnson grass that had been heated for six hours at 100°C. with moisture contents of from 2·0 per cent to 0·1 per cent, but the seedlings produced were less vigorous than their controls. Waggoner (1917) reported that 14 per cent of some seeds of the Icicle radish were viable after heating for thirty minutes at 123°C. with 0·4 per cent moisture content, but that all died which were heated at 100°C. with 4 per cent moisture, or at 65°C. with 40 per cent moisture. By pre-heating for one day at from 65° to 75°C., and for one day at 90°C., Dixon (1901) was able to germinate seeds of a number of plant species after heating them at temperatures ranging from 100° to 114°C. Atanasoff & Johnson (1920) heated well-dried seeds of wheat, barley, oats, and rye at 100°C. for fifteen and thirty hours, and of wheat and barley at from 100° to 110°C. for forty-five hours, and got good germination—except in the case of the rye seed, which germinated poorly. Earlier literature on heating seeds is reviewed in the papers of Waggoner (1917) and of Atanasoff & Johnson (1920).

Waggoner, and other investigators, have noted that heating of seeds may cause delay in their germination. Crocker & Groves (1915) found that wheat seeds delayed the beginning of their germination from four to fourteen days, depending upon the length of heating. Jozefowicz (1930) found that increasingly severe heat-treatments of tomato seeds caused progressive delay in their germination, which occurred from ten to eighteen days after planting. Seedlings grown from seeds that have been subjected to severe heat-treatments may show abnormalities of growth and form. Jozefowicz (1930) obtained from 50 to 100 per cent of abnormal tomato seedlings from seeds heated at 90° to 100°C. for one hour, unless the seeds were carefully pre-dried. Gain (1924) grew sunflower plants from desiccated embryos which had endured a series of successively more severe heat-treatments, ending with temperatures raised from 125° to 155°C. during about thirty minutes. The plants which were grown from embryos that had been so treated were very abnormal in form and growth, and although some of them produced flowers, none was able to form seeds (Gain, 1927).

A detailed experiment in which heat and moisture were investigated as factors in the increased mutation rate from *Datura* seeds, has been described by Cartledge *et al.* (1936). The seeds used were produced by self-pollinating *Datura* plants of a standard line which had been inbred since 1916, and had been passed through a haploid phase containing maternal chromosomes only. These seeds were subjected to treatments with various environmental factors, particularly with heat and controlled moisture content. After adjustment of the moisture content of the seeds to 2 to 15 per cent, they were placed in sealed vials and treated at temperatures of from 45° to 80°C. for from two hours to five days. The largest numbers were treated at 5 per cent moisture and 75° to 80°C. for from two to forty-eight hours. The most severe treatments killed the seeds, especially when these were high in moisture content, but moderate treatments increased the seedling yield over that of the controls. Good germination was obtained from treatments of seeds with 5 per cent moisture for twenty-four hours at 80°C., and for thirty-six hours at 75°C. The interval of time between planting and appearance of the seedlings was lengthened by increases in the temperature of treatment, by increases in moisture content of the seed, and especially by increases in duration of the treatment. Seeds with moderate and severe heat-treatments produced high percentages of plants with abnormal growth, as reflected in their types of branching.

The plants were tested for pollen-abortion mutations by microscopic examinations of pollen samples. Mutations of the pollen-abortion gene type were found in fifty-six plants, while pollen-abortion of the type caused by chromosomal mutations occurred in thirty-seven plants. Although a total of ninety-three pollen-abortion mutations were thus found in the 8,741 plants tested, none of these was found in the 920 control plants included. Most of the mutations involved a sector comprising half, or less than half, of the plant. Within the exposure times used, there was no significant increase of mutations at temperatures lower than 70°C. The highest mutation rates found were about 5 per cent, obtained from seeds with 5 per cent moisture content that had been heated at 75° and 80°C. Higher rates had been obtained from aged seeds (Cartledge & Blakeslee, 1934). In general, the mutation rate increased with increased temperature, with increased moisture content, and with increased duration of treatment—conditions which delayed the germination of treated seeds.

The primary purpose of the experiments by Cartledge et al. (1936) was to test the effects of treatments which had been selected as possible factors in the increased mutation rates obtained from aged seed.

The effect on the mutation rate of environmental factors such as heat and moisture, is of direct interest for the problem of the cause and nature of mutation. Although, not so many years ago, it was generally supposed that mutation was both rare in nature and beyond the reach of experimental stimulation, it is now well known that a variety of experimental procedures may be successfully employed to induce mutations. Those recognized at first were strong radiation treatments; but it has become increasingly apparent that treatments more nearly related to the normal environment of the organism are capable of producing similar, though less extreme, mutational effects. Heat is a factor in the normal environment, and one to which cells respond by quantitative or qualitative changes in their physiological processes. Navashin (1933) has suggested that age and heat treatments operate to produce mutations through alteration of the rate or character of the normal processes within the living cells. If this should be the case, then the mutations induced by age, heat, and the like, may be related in cause to those mutations considered natural or spontaneous. Increased mutation rates from ageing seeds of Datura (Cartledge & Blakeslee, 1934) may be explained as an accumulation of mutations occurring at a more or less constant rate. Stubbe (1935, 1935a) presents evidence for an accelerated rate of mutation with age in seeds of Antirrhinum. Heat, on the other hand, certainly increases the rate of mutation; but the effects of age and of heat may be reducible to a time-temperature relationship which will include both treatments.

It is of some interest to observe that the experimental determinations of normal mutation rates are usually made under carefully controlled conditions with fresh material. The increases of mutation associated with ageing, high moisture content, and heating of the seeds, suggest that conditions beyond the limits of carefully controlled experiments may be of importance in determining mutation rates in nature, and that these rates may be higher than the 'normal' rates found by experimental tests.

It is obvious from the results of Cartledge et al. (1936) that the maximum mutation rate precedes the death of the seeds, as the treatments giving the highest pollen-abortion rates still permitted excellent germination. This was also noted by Navashin et al. (1940), who found that the life of the seed of Crepis was proportional to the time of storage, but the mutation rate was proportional to the time only in the early storage period, i.e. the first 1,000 days. After 1,000 days a $1\frac{1}{2}$-fold increase in the storage period resulted in nearly nine times more mutations.

More recent work has served to emphasize the increased mutation rate with ageing of different seeds. Mutants characterized by abnormal height and increased number of leaves appeared as a result of ageing seeds of *Nicotiana tabacum* (Gisquet *et al.*, 1951). The roots of lettuce and onion seedlings developed from old seeds show much chromosome breakage, resulting in bridges and fragments at anaphase (Lewis, 1953). Harrison & McLeish (1954), working also with lettuce and onion roots, found a general trend to a sharp decrease in chromosome abnormalities at high germination values, and a corresponding rise in mitotic activity. Results with onion, however, revealed a low incidence of chromosome breakage at all levels of germination, showing that loss of viability is not necessarily determined by, nor does it precede, increase in the frequency of abnormal cells. These authors conclude (Harrison & McLeish, 1954) that the production of cytological abnormalities must be a consequence rather than the initial cause of seed deterioration.

Cytological and genetical changes occurring in ageing seeds of common and durum wheat, barley, rye, and peas, that had been stored from one to thirty-two years, have been studied by Gunthardt *et al.* (1953). Their analysis included a record of (1) chromosomal bridges and fragments in mitotic anaphase of root-tip and shoot-tip cells of seeds of various ages, (2) chromosomal interchanges in the pollen-mother cells of plants grown from these seeds, and (3) seedling mutations in the progeny of these seeds. Examples of chromosomal aberrations are shown in Plate 13 (Gunthardt *et al.*, 1953). Both chromosomal aberrations and genetic mutations are described, and the types of the former appear to be identical to those arising from ionizing radiations. The authors also show that the dosage of natural radiation, including cosmic radiation, is not sufficient to account for the changes taking place in the ageing seeds.

Gustafsson (1937), in his description of death as a nuclear process, says that vital processes are going on in the nuclei of the embryo in the seed preparatory to the reproduction of chromosomes and nuclei. If these processes have advanced to a certain stage, but external conditions prevent their fulfilment in the actual division, the nuclei degenerate and die. Increased moisture content hastens this stage, and thus shortens the life of the seed.

Amato (1954) has critically reviewed much of the literature on physiological and genetic aspects of seed ageing in connection with the problem of spontaneous mutability in plants, and has given an extensive bibliography on the subject.

As storage conditions can overcome the effect of ageing, as measured by years of life, a new definition for ageing in seeds is needed. Longevity is more closely related to the morphological and physiological state of the seed, especially of the embryo, than to the length of the storage period. 'Ageing' takes place in a few hours under some conditions, while under optimum conditions ageing might require centuries. More basic research is needed on this problem, using such new 'tools' as chromatography and radioactive tracers, to determine the basic causes of loss of viability.

XVI

PRACTICAL CONSIDERATIONS

TRANSPORTATION

THERE is a good chance that seeds, especially those which are very sensitive to high humidity and temperatures, will lose their viability in transit. Large consignments of field seeds could be killed by overheating in railway trucks or the holds of ships. The first thought is that such deterioration could be prevented by drying the seeds and dispatching them in sealed containers. This would probably be effective for most seeds, but there still remain some which do not tolerate desiccation and must be kept moist on the way. This is a difficulty in transporting palm seeds. However, a satisfactory method of packing palm seed for distant transportation has been developed at the Puerto Rico Experiment Station (Anon., 1940b). Experience at the station suggested that the failure of palm seeds to germinate may have been the result of improper packing of the seeds at their source. As the seeds must be kept moist, they are often made too wet and hence germinate, the resultant seedlings dying in transit. Excellent results can be obtained by packing the seeds in a medium of granulated peat-moss or the dusty, granular, corky material from coir, in either case having a moisture content of approximately 30 per cent. The mixture of seeds and packing material should be hermetically sealed in ordinary tin cans. If the proper moisture content has been obtained for the packing material, and the tin container is properly sealed, the palm seeds will not lose sufficient moisture to lower their viability, while on the other hand they cannot absorb sufficient moisture to germinate during transit.

This problem has been overcome in a different way by Kurokami *et al.* (1947). It has been the practice to pack seeds of chestnut, which are killed by drying, in moistened sawdust for storage or carriage. This is a good plan for storage at low temperatures, for germination does not take place or is, at least, delayed. However, during storage at high temperatures, such as would be encountered in transit, there is danger of germination and subsequent loss. These authors inhibited such germination by using 2 to 4 gm. of the phytohormone, naphthalene potassium acetate, or its methyl ester, per packing box. The boxes were approximately 1 ft. 8 in. long, 1 ft. wide, and 1 ft. deep. Sawdust layers at the top and bottom were about 4 in. in thickness. The phytohormones were prepared in aqueous

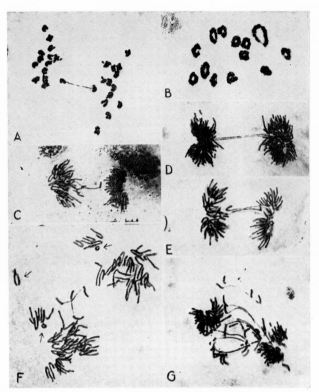

Photo. by courtesy Washington Agricultural Experiment Station, Pullman,
Washington, U.S.A.

PLATE 13

Examples of chromosomal aberrations occurring in plants from
aged wheat seeds. A, Kubanka 1919: Bridge in mitotic anaphase,
× 900. B, Kubanka 1919: Ring-of-four in mitotic metaphase, × 900.
C, Kubanka 1919: Bridge locked into a dicentric chromosome, × 1080.
D, Baart 1927: Two parallel bridges in mitotic anaphase, × 1080.
E, Kubanka 1919: Two interlocked bridges in mitotic anaphase,
× 1080. F, Baart 1925: Two crossed bridges, one pair of acentric rod
fragments, and two acentric rings, in mitotic anaphase, × 1080.
G, Baart 1925: Nine bridges and several fragments (some paired) in
mitotic anaphase, × 1080. (From Gunthardt, Smith, Haferkamp &
Nilan, 1953.)

solution, which was sprinkled over the sawdust with a watering can and then mixed into the sawdust by hand. Depending on the market price of plant hormones, this method can be put to practical use in the storage of chestnut seeds over extended periods of time, or in the transport of these seeds over long distances through sub-tropical regions (Kurokami *et al.*, 1947).

In 1941, a ton of selected rubber seed was flown to Brazil in three large army bombers (Anon., 1941*b*). The seed was collected from high-yielding clones grown in the Philippines, and was shipped to the Canal Zone by boat. Delay in reaching the Canal Zone made it imperative to deliver the seeds quickly to prevent deterioration. For small quantities of valuable seeds, the aeroplane offers a method of maintaining viability in transportation, but it is not yet practical for large seed-lots.

It is the opinion of Degen & Puttemans (1931) that seeds do not lose their germination power as a result of shipment across the ocean or the Equator, but rather in storage in tropical countries. Farmers and horticulturists living in tropical countries and buying their seeds in Europe, should secure them as short a time as possible before they are to be used. These two authors, Degen in Brazil and Puttemans in Hungary, exchanged samples of seeds from different plant families and variously sensitive to climatic changes, before arriving at the above conclusions.

Clearly some seeds are killed by transit conditions while others are not. More knowledge is needed of the tolerances of various seeds to transport conditions.

Storage Facilities

Governments of various countries, in addition to individuals in responsible agricultural positions, are becoming increasingly aware of the possibilities of retaining both food and seeding value of seeds by proper storage.

In India, for example, rice grain is more subject to deterioration than that of any other cereal crop, for it is grown and stored under humid and often primitive conditions (Panikkar, 1947). About 317,000 tons, 1·5 per cent of the annual crop, is lost in storage—a loss important to a people living near the subsistence level. India also must import a large quantity of vegetable seeds each year, because of the short life-span of such seeds in the Indian climate (Dutt & Thakurta, 1953–55). More thought and experimental data are being applied to a practical solution of this problem, so that surplus stocks of vegetable seeds will not have to be wasted for lack of proper storage facilities.

A survey has been made of the problem associated with the storage of grain and the methods used in Central and East Africa, with a view to recommendations of means to prevent storage losses (Oxley, 1950).

Although the best condition for maintaining viability is obtained through control of both temperature and humidity, it has been found that seed can be preserved for a reasonable period at atmospheric temperature in Puerto Rico, provided the relative humidity is reduced to 20 per cent or less (Hopkins *et al.*, 1947). A practical and cheap dehydrating agent was discovered in ordinary clay subsoil which had been oven-dried to remove the hygroscopic moisture. As a result of these findings, practical storage methods can be recommended.

For commercial purposes in cooler, less humid climates such as those of New South Wales (Myers, 1943) or Denmark (Dorph-Petersen, 1928), no special storage is necessary if the seeds are to be kept for a short period, i.e. one or two years.

A cold-storage room is an excellent place to store large quantities of seeds, according to Akamine (1943), though care must be taken not to allow the drip from melting ice to come in contact with the seeds. The seeds should be placed in the driest part of the cold-storage room. Other work has shown the danger of a cold-storage room with a humid atmosphere, unless the temperature is below freezing-point.

Weibull's data (Weibull, 1955) also show the possibility of using a commercial cold-storage facility, but he recommends an operating temperature of −20°C. for long-term storage of certain vegetable and flower seeds. He points out that, with cold storage, the seed trade could be made less risky and the supply of seeds more regular; prices could be kept at a more even level, and the plant breeder could concentrate more efficiently upon his stock-seed production and accumulate bigger stock-seed reserves.

In the United States as of 1954 (Anon., 1954), storage losses of cereals (wheat, oats, barley, rye, and rice grain) amounted to the harvest from 3,676,000 acres, valued at $89,666,000, and representing a 4·5 per cent loss. This was exclusive of losses due to insects, which accounted for a loss of the production from an additional 1,428,000 acres, representing a loss of 2·4 per cent of the crop. Recommendations for practical drying procedures and storage facilities have been made for specific crop seeds in the various states; i.e. in Texas for rice (Hildreth & Sorenson, 1957), in Illinois for soya beans (D. G. Carter & Holman, 1952), in Kansas for wheat (J. L. Schmidt, 1955), and in South Carolina for cotton (Brand & Sherman, 1913). Nebraska has provided a source of pure seed both of newly developed varieties and of the more important established varieties (Sahs, 1957). The federal soil conservation service is providing nursery storage cellars at Pullman, Washington, with air conditioning, to help preserve tree seeds over long periods of time (Anon., 1940). A national seed storage facility where seed stocks will be preserved for future use has been constructed on the campus of Colorado State University at Fort Collins, Colorado (Anon., 1958; Binkley, 1958). It will serve as a federal germ-plasm bank on which plant breeders can draw for work with thousands of different plants representing the world's most valuable food, feed, pasture, fibre, and tree crops. It is hoped that this will prevent the loss of many new desirable forms as well as old varieties during years of active plant breeding. Seed stock from all over the world will be stored for future reference, and a seed-testing laboratory will be part of this new building.

Descriptions of seed storage rooms, practices, and equipment, as employed in various localities, are to be found in the following articles: Anon., 1931; Long & Cropsey, 1941; Cartter, 1942; Eggleston, 1949; Singley et al., 1954; Swaine, 1954; and Pingale & Balu, 1955.

Recommendations such as have been outlined in this chapter could not have been made without numerous experiments to determine the specific requirements for safe storage of various kinds of seeds. They represent a great advance in seed technology. At least two experiments are now in progress to obtain detailed information on seeds over longer storage periods. One of these has been described by Went (1948) to last a possible 360 years. The other involves lettuce, onion, and tomato seeds, in sufficient quantity for 150 years of testing, which have been stored at the Boyce Thompson Institute for Plant Research, Inc., Yonkers, New York (unpublished). As, in this latter instance, dry seeds are stored in sealed containers at a temperature of $-17.8°C.$, it may well be that some of the seeds will still be viable at the end of 150 years.

GLOSSARY

Accessory fruit. One in which some other part of the plant not belonging to the pistil (gynaecium) is associated with the matured ovary.

Achene. A small, dry, one-seeded indehiscent fruit.

Actinomycytes. A group of filamentous organisms related to bacteria.

Aerobic. Living or active only in the presence of oxygen.

Aggregate fruit. A fruit formed from several ovaries produced by one flower.

Alkaloid. An organic substance occurring naturally in plants, e.g. morphine.

Amylase. An enzyme accelerating decomposition of starch.

Analysis of variance. A statistical treatment of experimental results.

Anaphase. A stage in cell division during divergence of daughter chromosomes.

Anatomy. The science which treats of structure or, more widely, of form. **Plant anatomy** deals with the cells which make up plants.

Angiosperms. Plants with seeds borne within the ovary.

Annual. Of only one year's duration, or less.

Anther. The pollen-bearing portion of a stamen.

Aquatic. Living in water.

Ascorbic acid. Pure Vitamin C.

Auxin. Growth-regulating hormone.

Biochemistry. The chemistry of living organisms and their components and products.

Calyx. The outer whorl of floral 'leaves'.

Carbohydrates. Starches and sugars.

Carpel. A simple pistil or one member of a compound pistil.

Catalase. An enzyme which decomposes hydrogen peroxide into water and oxygen.

Chromosome. A deeply-staining body, normally of a constant number for one species, found in the nucleus during cell division.

Clone. A group of plants derived from a single plant by vegetative propagation or artificial separation.

Cotyledon. The foliar portion or first leaves (one, two, or occasionally more) of the embryo as found in the seed.

Critical moisture content. Seed moisture content above which deterioration is rapid.

Crude fibre. Material obtained as residue in the chemical analysis of plant substances.

Cuticle. The thin, usually waxy covering of the outside of epidermal cells of most land plants.

Cytological. Pertaining to the cell.

Dehiscent. Opening spontaneously along certain lines or in a definite direction.

Dehydrogenase. An enzyme accelerating the removal of hydrogen from tissues.

Desiccation. Drying out.

Dextrin. A soluble substance derived from starch.

Diaphorase. A dehydrogenase containing riboflavin.

Diastase. An enzyme which acts principally in converting starch into sugar.

Dicotyledons. Plants having seeds with two cotyledons or seed-leaves, and comprising a group including most of our familiar broad-leafed trees and vegetables, etc.

Drupe. A fleshy fruit with the inner portion of the pericarp hard or stony.

Electrolyte. A conductor of electricity.

Electromotive force. One causing electrical action or effects.

Embryo. The rudimentary plantlet within the seed.

Endosperm. The nutritive tissue of certain seeds.

Enzyme. A chemical acting on one or more specific substrates; a ferment.

Epidermis. The outside layer of cells of the aerial parts of higher plants.

Ferment. An enzyme.

Fertilization. Fusion of male and female nuclei to form the beginnings of a new individual.

Fluorescence. The property of emitting radiation.

Fruit. The seed-bearing portion of the plant, formed from the ripened ovary together with other plant parts which may be associated with it.

Fungi. A great group of lower plants without chlorophyll, and typically saprophytic or parasitic. They cause many plant diseases.

Fungicide. A substance which destroys fungi.

Galvanometer. An instrument for measuring a small electric current.

Gene. A unit hereditary factor in the chromosome.

Genetics. The branch of biology dealing with heredity and variation.

Grain. The fruit of cereals.

Gymnosperms. Plants bearing naked seeds without an ovary.

Haploid. Organism (or state of) having the number of chromosomes characteristic of one mature germ-cell.

Hemicellulose. One of several polysaccharides occurring as constituents of cell-walls, cotyledons, endosperms, and woody tissues.

Hermetically sealed. Air-tight.

Hesperidium. A berry covered with a tough rind, as in orange.

Homozygous. Having identical genes for a given character.

Hybrid. A cross-breed of two species (normally).

Hydrolysis. Reaction between a compound and water.

Hygroscopic. Sensitive to or retaining moisture.

Hypocotyl. Part of the embryo between the cotyledon and the radical in the seed.

Imbibed seeds. Seeds which have absorbed liquid.

Indehiscent. Not opening or splitting at maturity.

Insecticide. A substance which kills insects.

in vitro. In a test-tube; outside of the organism.

Jacobsen germinator. A special device for seed testing in which water is supplied to a blotting-paper substrate by means of wicks.

Latin square. An experimental design containing the same number of replications and treatments.

Lignification. Thickening of plant cell-walls by deposition of wood.

Longevity. Life-span.

Metabolism. The chemical changes, constructive and destructive, occurring in living organisms.

Microflora. Microscopic plants, such as fungi.

Micro-organisms. Microscopic organisms.

Mitosis. Nuclear division.

Monocotyledons. Plants having seeds with one cotyledon or seed-leaf, and forming a group (including grasses, palms, etc.) that with the dicotyledons (*q.v.*) make up the Angiosperms (*q.v.*).

Monosaccharide. A simple sugar, not decomposed by hydrolysis.

Morphology. The science of (mostly gross external) form and structure of living organisms.

Multiple fruit. One resulting from the union or compact aggregation of ovaries of more than one flower.

Mutation. Variation producing a definite, inherent and heritable change in an organism.

Necrotic tissues. Dead tissues.

Nicotinic acid. The pellagra-preventive factor of Vitamin B complex.

Nucleus. The complex and usually spheroidal mass essential to the life of most cells.

Nut. A hard, indehiscent, one-seeded fruit.

Organic acid. An acid containing at least the elements carbon, oxygen, and hydrogen, and normally produced by or derived from tissues of living organisms.

Ovary. The part of the pistil (gynaecium) which contains the ovules.

Ovule. The body which becomes the seed after fertilization.

Oxidase. An enzyme which promotes oxidation.

Palisade cells. An arrangement of elongated cells to form a tissue, usually constituting the uppermost green layer in the leaf.

Pathogen. An organism causing a disease.

Pepo. A fleshy fruit with a hard rind, e.g. a squash or melon.

Perennial. Persisting for a number of years.

Pericarp. The matured ovary-wall.

Permeability. State of being penetrable.

Peroxidase. An enzyme which causes activation of peroxide oxygen.

Phenolase. An enzyme that oxidizes phenols.

Physiology. The science of function and activity of organisms.

Phytohormones. Internal secretions of plants; plant hormones.

Pistil (Gynaecium). The seed-bearing organ or organs of the flower consisting of ovary, style (sometimes absent), and stigma.

Plasmolysis. Withdrawal of water from plant cells, causing a contraction of cell contents.

Plumule. The part of the embryo which forms the shoot of the plant.

Pollen. The 'male' grains contained in the anther.

Pollination. The process of depositing pollen upon the stigma.

Polyethylene. A plastic film.

Polyploid. A plant with more than two sets of chromosomes.

Polysaccharide. A carbohydrate decomposable by hydrolysis into two or more simple sugars or monosaccharides.

Pome. An indehiscent, two- or more-celled fleshy fruit, as in apple.

Protein. Albuminous substance.

Proteolytic enzymes. Those which bring about decomposition of protein.

Radicle. The portion of the embryo which forms the root of the new plant.

Radiocarbon dating. A method of age-determination based on the amount of radioactive carbon found in the sample, as compared with the amount of such carbon known to be present in young tissues of the same type.

Radiograph. An instrument for measuring and recording radiation.

Receptacle. The more or less expanded portion of a plant bearing the flower-parts.

Reducing sugar. A mono- or di-saccharide that reduces (copper or silver salts) in alkaline solution.

Relative humidity. Ratio of the quantity of vapour actually present in the air to the greatest amount it can contain at a given temperature.

Respiration. Gaseous interchange between an organism and its surrounding medium; normally the taking in of oxygen and the giving out of carbon dioxide.

Riboflavin. Vitamin B_2.

Samara. An indehiscent, winged fruit, as in ash.

Saprophyte. A plant which lives on dead or decaying organic matter.

Scutellum. Cotyledon peculiar to the grasses, being a shield-shaped organ through which the embryo absorbs food from the endosperm.

Seed. The ripened ovule, consisting of the embryo and coats, with often some additional stored food.

Silica gel. A form of colloidal silica possessing many fine pores, and hence very absorbent.

Simple fruit. One resulting from the ripening of a single ovary.

Spermatophyte. Plant producing seeds; a seed-plant.

Stratification. Low-temperature pretreatment of seeds—originally given with layers of seeds alternating with layers of sand or soil.

Style. The usually elongated portion of the pistil (gynaecium) connecting the stigma and ovary.

Substrate. A substance upon which something grows or an enzyme acts.

Sucrose. Cane sugar.

Taxonomy. The science of plant and animal classification.

Thiamin. Vitamin B_1.

Topographic. Descriptive of a place.

Vapour pressure. The pressure of a confined body of vapour.

Viability. The condition of being viable in the sense of being capable of growth and survival.

Vinyl film. A special type of plastic.

Vitality. Speed and energy of growth.

Vitamin. Accessory food factor.

BIBLIOGRAPHY*

ABBOTT, E. V. (1950) Some observations on sugarcane flowering and seed production in Louisiana and an experiment on production and storage of true sugarcane seed. *Sug. Bull. N. Orleans*, 28(21), 329–32. (Abstr. in *Biol. Abstr.*, 25, No. 25006, 1951.)

ÅBERG, E. (1950). Barley and Wheat from the Saqqara pyramid in Egypt. *Lantbr.-Hoögsk. Ann.*, 17, 59–63. (Abstr. in *Biol. Abstr.*, 25, No. 26373, 1951.)

ACTON, E. H. (1893). Changes in the reserve materials of wheat on keeping. *Ann. Bot., Lond.*, 7, 383–7.

ADAMS, J. D., NILAN, R. A., & GUNTHARDT, H. M. (1955). After-effects of ionizing radiation in Barley. I: Modification by storage of X-rayed seeds in oxygen and nitrogen. *Northw. Sci.*, 29, 101–8. (Abstr. in *Nuclear Sci. Abstr.*, 9, No. 7632, 1955.)

AFANASIEV, M. (1943). Germinating *Nandina domestica* seeds. *Amer. Nurserym.*, 78(9), 5–6.

AHLGREN, G. H., FISKE, J. G., & DOTZENKO, A. (1950). Viability of bromegrass seed as affected by dehulling and by storage in fertilizer. *Agron. J.*, 42, 336–7.

AKAMINE, E. K. (1943). The effect of temperature and humidity on viability of stored seeds in Hawaii. *Bull. Hawaii Agric. Exp. Sta.*, 90, 23 pp.

—— (1951). Viability of Hawaiian forest tree seeds in storage at various temperatures and relative humidities. *Pacif. Sci.*, 5, 36–46.

—— & RIPPERTON, J. C. (1938). Germination and viability of crop seeds in Hawaii. *Rep. Hawaii Agric. Exp. Sta.*, 1937, 10–12.

AKI, S. & WATANABE, S. (1955). Studies on the storage of the seeds of Kintoki carrot. I: Influence of storage temperature and seed moisture on the germination of seeds. *Kagawa Kenritsu Nôka Daigaku, Gakujutsu Hôkoku (Tech. Bull. Kagawa Agric. Coll., Japan)*, 7(1), 31–35. (In Japanese, with English summary.)

ALBO, G. (1908). Les enzymes et la faculté germinative des graines. *Bibl. Univ. Arch. Sci. Phys. et Nat. Ser.*, 4 25, 45–52.

ALLEN, G. S. (1957). Storage behavior of conifer seeds in sealed containers held at 0°F., 32°F., and room temperature. *J. For.*, 55, 278–81.

ALTSCHUL, A. M. (1949). Sweet and unspoiled. Chemical treatment prevents deterioration of cottonseed under storage. *Chem. Ind. Phil.*, 64 (2), 211.

—— CONDON, M. Z. & LAMBOU, M. G. (1951). Method of prevention of deterioration in seeds. U.S. Pat., No. 2,571,095.

AMATO, F. D' (1954). Di alcuni fisiologici e genetici dell' invechiamento dei semi. Contributo al problema della senescenza e della mutabilità spontanea nei vegetali. [On some physiological and genetical aspects of seed ageing. A contribution to the problem of ageing and of spontaneous mutability in plants.] *Caryologia*, 6(2) 217–40. (In Italian only; abstr. in *Biol. Abstr.*, 29, No. 29695, 1955.)

ANDERSON, J. A. & ALCOCK, A. W. (Ed.) (1954). *Storage of cereal grains and their products. Monog. Ser. Amer. Ass. Cereal Chem.*, vol. II, 515 pp.

ANDRÉN, F. (1945). Lagringsförsök med betat utsäde. [Storage experiments with disinfected seed.] *Växtskyddsnotiser,, Stockh.* 1945 (1), 1–6. (In Swedish only; abstr. in *Rev. Appl. Mycol.*, 25, 207, 1946.)

*The citation of an abstract indicates that the original publication was not seen.

ANON. (1843). Mummy wheat. *Gdnrs'. Chron.*, 1843, 787–8.

—— (1928). Noble fir seed keep well in cold storage. *For. Wrkr., U.S.*, 4(4), 14–15.

—— (1931). Regulations for warehousemen storing seeds under the United States warehouse act. *U.S. Dep. Agr. Bur. Agric. Econ. S. & R.A.*, No. 122, pp. 21.

—— (1932). Influence of age of pumpkin seed on yield. *Qd. Agric. J.*, 37, 74.

—— (1933). Samen aus mumien. *Mitt. Dtsch. Dendrol. Ges.*, 45, 345.

—— (1934). Mummy wheat. *Nature, Lond.*, 134, 730.

—— (1935). Valuable information for seedsmen (Bodger Seeds, Ltd., El Monte, Cal.). *Flor. Exch.*, 85(1), 5.

—— (1936). Connecticut Agricultural Experiment Station. Report of the director for the year ending October 31, 1935. *Bull. Conn. Agric. Exp. Sta.*, 381, 165–202.

—— (1940). Air-condition seeds. *Amer. Nurserym.*, 72(5), 31–32.

—— (1940a). *Cinchona* seeds lose viability rapidly. *Rep. P.R. Agric. Exp. Sta.*, 1940, 18–19.

—— (1940b). A satisfactory method of packing palm seed for distant transportation was developed. *Rep. P.R. Agric. Exp. Sta.*, 1939–40, 62–63.

—— (1941). Storage of red pine seed. *Tech. Notes Lake St. For. Exp. Sta.*, 1 p.

—— (1941a). Storing coffee seed. *Trop. Agriculture, Trin.*, 18(2), 25.

—— (1941b). Planes taking rubber seed from Canal Zone to Brazil. *Chron. Bot.*, 6(14), 325.

—— (1941c). Viability studies on tree and shrub seed. *Rep. N.Y. St. Agric. Exp. Sta.*, 1939–40, p. 41.

—— (1942). Duration of viability in seeds. *Gdnrs'. Chron.*, 111(2893), 234. (Abstr. in *Biol. Abstr.*, 17, No. 6934, 1943.)

—— (1945). Good bread made from twenty-year-old wheat. *Sci. News Lett., Wash.*, 47, 184.

—— (1947). Test seeds by electricity. *Amer. Nurserym.*, 85(12), 6.

—— (1948). Woody-plant seed manual. *Misc. Publ. U.S. Dep. Agric.*, 654, 290–2, 361–3.

—— (1949). Dry cold storage of seed corn. *U.S. Dep. Agric. Rep. Agric. Exp. Sta.*, 1948, 43–44.

—— (1950). Seed corn keeps best when stored near zero in tight, dry container. *Crops & Soils*, 2(9), 29.

—— (1951). Ancient seeds not viable. *Sci. News Lett., Wash.*, 59, 180.

—— (1951a). Deep freezing prolongs viability of sugarcane seed. *U.S. Dep. Agric. Rep. Administrator of Agric. Res.*, 1951, 379.

—— (1952). Low temperature storage keeps sugarcane seed viable many months. *Crops & Soils*, 4(8), 27.

—— (1952a). *Manual for testing agricultural and vegetable seeds. Agric. Handb. U.S. Dep. Agric.*, No. 30, Washington, D.C., 440 pp.

—— (1953). Longevity of *Halogeton* seeds. *Mon. Bull. Calif. Dep. Agric.*, 42(4), 290–1.

—— (1953a). Recent research on drying and storage of rough rice. *Ser. Bull. Southern Co-op.*, 29, 29 pp.

—— (1954). Losses in agriculture. *Res. Ser. U.S. Dep. Agric.*, 20(1), 95–101.

—— (1954a). The preservation of viability and vigor in vegetable seed. *ASGROW Monogr.*, 2, 32 pp.

—— (1956). International rules for seed testing. *Proc. Int. Seed Test. Ass.*, 21, 1–80.

—— (1958). News Briefs. *Science*, 128, 1563.

ANTHONY, K. R. M. & TARR, S. A. J. (1952). The causes of deterioration of cotton seed in the equatorial province of the Anglo-Egyptian Sudan. *Emp. J. Exp. Agric.*, 20(77), 56–65.

ARAI, M. & KATAOKA, T. (1956). Ecological studies on *Alopecurus aequalis* Sobol. (3) Influence of soil moisture on the dormancy and longevity of seeds. (4) Seasonal variation in the viable seed population and its vertical distribution in the soil. *Proc. Crop Sci. Soc. Japan*, 24(4), 319–23. [In Japanese, with English summary, p. 323.]

ARMOLIK, N. & DICKSON, J. G. (1956). Minimum humidity requirement for germination of conidia of fungi associated with storage of grain. *Phytopathology*, 46, 462–5.

—— —— & DICKSON, A. D. (1956). Deterioration of barley in storage by microorganisms. *Phytopathology*, 46, 457–61.

ARNDT, C. H. (1946). The internal infection of cotton-seed and the loss of viability in storage. *Phytopathology*, 36, 30–37.

ATANASOFF, D. & JOHNSON, A. G. (1920). Treatment of cereal seeds by dry heat. *J. Agric. Res.*, 18, 379–90.

ATKINS, W. R. G. (1909). The absorption of water by seeds. *Sci. Proc. R. Dublin Soc.* [N.S.], 12, 35–46. (Abstr. in *Exp. Sta. Rec.*, 21, 725–6, 1909.)

AVERY, A. G. & BLAKESLEE, A. F. (1936). Visible mutations from aged seeds (Abstract). *Amer. Nat.*, 70, 36–37.

BAILEY, C. H. (1917). The moisture content of heating wheat. *J. Amer. Soc. Agron.*, 9, 248–51.

—— (1917a). The handling and storage of spring wheat. *J. Amer. Soc. Agron.*, 9, 275–81.

BAILEY, S. W. (1955). Air-tight storage of grain; its effects on insect pests. I: *Calandra granaria* L. (Coleoptera, Curculionidae). *Aust. J. Agric. Res.*, 6(1), 33–51.

—— (1956). Air-tight storage of grain; its effects on insect pests. II: *Calandra oryzae* (small strain). *Aust. J. Agric. Res.*, 7(1), 7–19.

BAIRD, P. D., MACMASTER, M. M. & RIST, C. E. (1950). Studies on a rapid test for the viability of corn for industrial use. *Cereal Chem.*, 27, 508–13.

BAKER, D., NEUSTADT, M. H. & ZELENY, L. (1957). Application of the fat acidity test as an index of grain deterioration. *Cereal Chem.*, 34, 226–33.

BAKKE, A. L. & NOECKER, N. L. (1933). The relation of moisture to respiration and heating in stored oats. *Bull. Ia. Agric. Exp. Sta.*, 165, 317–36.

BALDWIN, H. I. (1934). Germination of red spruce. *Plant Physiol.*, 9, 491–532.

—— (1935). Catalase activity as a measure of viability of tree seeds. *Amer. J. Bot.*, 22, 635–44.

—— (1942). *Forest tree seed of the north temperate regions with special reference to North America.* Chronica Botanica, Waltham, Mass., U.S.A., 240 pp.

—— & FLEMION, F. (1941). Rapid method for testing white pine germination. *Fox For. Notes*, No. 30.

BALL, G. J. (1935). Serious losses in germination of seed can generally be avoided. *Flor. Rev.*, 75(1937), 19–20, 49–51.

BARNER, H. & DALSKOV, F. (1955). Notes on storage of Douglas fir seed. *Proc. Int. Seed Test. Ass.*, 20, 57–61.

BARRONS, K. C. & MCLEAN, D. M. (1945). A study of the causes of low germination of radish seed crops. *Quart. Bull. Mich. Agric. Exp. Sta.*, 27, 398–408.

BARTLETT, L. (1859). Vitality of seeds. *Report of the Commissioner of Patents for the Year 1858, Washington, D.C.*, pp. 332–3.

BARTON, L. V. (1930). Hastening the germination of some coniferous seeds. *Amer. J. Bot.*, 17, 88–115. (*Contr. Boyce Thompson Inst.*, 2, 315–42, 1930.)

—— (1932). Effect of storage on the vitality of delphinium seeds. *Contr. Boyce Thompson Inst.*, 4, 141–54.

—— (1933). Germination and storage of delphinium seed. *Bull. Amer. Delphinium Soc.*, 2(1), 12–14. (*Prof. Pap. Boyce Thompson Inst.*, 1(26), 248–50, 1933).

—— (1935). Storage of vegetable seeds. *Contr. Boyce Thompson Inst.*, 7, 323–32.

—— (1935a). Storage of some coniferous seeds. *Contr. Boyce Thompson Inst.*, 7, 379–404.

BARTON, L. V. (1939). A further report on the storage of vegetable seeds. *Contr. Boyce Thompson Inst.*, **10**, 205–20.

—— (1939*a*). Storage of elm seeds. *Contr. Boyce Thompson Inst.*, **10**, 221–33.

—— (1939*b*). Storage of some flower seeds. *Contr. Boyce Thompson Inst.*, **10**, 399–427.

—— (1939*c*). Germination and storage of lily seeds. *Nat. Hort. Mag.*, **18**, 193–4.

—— (1939*d*). Experiments at Boyce Thompson Institute on germination and dormancy in seeds. *Sci. Hort.*, **7**, 186–93.

—— (1941). Relation of certain air temperatures and humidities to viability of seeds. *Contr. Boyce Thompson Inst.*, **12**, 85–102.

—— (1943). Effect of moisture fluctuations on the viability of seeds in storage. *Contr. Boyce Thompson Inst.*, **13**, 35–46.

—— (1943*a*). The storage of citrus seeds. *Contr. Boyce Thompson Inst.*, **13**, 47–55.

—— (1945). A note on the viability of seeds of Maga, *Montezuma speciosissima*. *Contr. Boyce Thompson Inst.*, **13**, 423–6.

—— (1945*a*). Viability of seeds of *Fraxinus* after storage. *Contr. Boyce Thompson Inst.*, **13**, 427–32.

—— (1945*b*). Respiration and germination studies of seeds in moist storage. *Ann. N.Y. Acad. Sci.*, **46**, 185–208.

—— (1947). Effect of different storage conditions on the germination of seeds of Cinchona Ledgeriana Moens. *Contr. Boyce Thompson Inst.*, **15**, 1–10.

—— (1948). Storage of seeds of the Regal Lily. *Boyce Thompson Inst. Prof. Paper*, **2**(6), 45–51.

—— (1949). Seed packets and onion seed viability. *Contr. Boyce Thompson Inst.*, **15**, 341–52.

—— (1953). Dormancy in seeds. *Rep. 13th Int. Hort. Congr., Lond.*, 1952, **2**, 1001–12.

—— (1953*a*). Seed storage and viability. *Contr. Boyce Thompson Inst.*, **17**, 87–103.

—— (1954). Effect of subfreezing temperatures on viability of conifer seeds in storage. *Contr. Boyce Thompson Inst.*, **18**, 21–24.

—— (1954*a*). Storage and packeting of seeds of Douglas fir and Western hemlock. *Contr. Boyce Thompson Inst.*, **18**, 25–37.

—— & CROCKER, W. (1948). *Twenty years of seed research at Boyce Thompson Institute for Plant Research.* Faber & Faber, London, 148 pp.

—— & GARMAN, H. R. (1946). Effect of age and storage condition of seeds on the yields of certain plants. *Contr. Boyce Thompson Inst.*, **14**, 243–55.

—— & HOTCHKISS, J. E. (1951). Germination of seeds of *Eichhornia crassipes* Solms. *Contr. Boyce Thompson Inst.*, **16**, 215–20.

BARTON-WRIGHT—*see* WRIGHT, E. C. BARTON-.

BATES, C. G. (1930). One-year storage of white pine seed. *J. For.*, **28**, 571–2.

BATTLE, W. R. (1948). Effect of scarification on longevity of alfalfa seed. *J. Amer. Soc. Agron.*, **40**, 758–9.

BAYFIELD, E. G. & O'DONNELL, W. W. (1945). Observations on thiamin content of stored wheat. *Food Res.*, **10** (6), 485–8. (Abstr. in *Biol. Abstr.*, **20**, No. 7183, 1946.)

BEATTIE, J. H. (1931). Peanut seed may be kept for several years under proper conditions. *Yearb. Agric. U.S. Dep. Agric.*, 1931, 426–7.

—— & BOSWELL, V. R. (1938). Longevity of onion seed in relation to storage conditions. *Proc. Amer. Soc. Hort. Sci.*, **35**(1937), 553.

—— —— (1939). Longevity of onion seed in relation to storage conditions. *Agric. Circ. U.S. Dep.*, **512**, 23 pp.

—— JACKSON, A. M. & CURRIN, R. E. (1932). Effect of cold storage and age of seed on germination and yield of peanuts. *Agric. Circ. U.S. Dep.*, **233**, 12 pp.

—— & KUSHMAN, L. J. (1947). Effect of moisture contents of peanuts on shelling damage. *Peanut J. and Nut World*, **27**(1), 35. (Abstr. in *Biol. Abstr.*, **23**, No. 16431, 1949.)

BEATTIE, J. H. & TATMAN, E. C. (1950). Long-term storage of parsnip seed. *Proc. Amer. Soc. Hort. Sci.*, 55, 435–7.

BECQUEREL, P. (1904). Sur la perméabilité aux gaz de l'atmosphère, du tégument de certaines graines desséchées. *C.R. Acad. Sci., Paris*, 138, 1347–9.

—— (1907). Recherches sur la vie latente des graines. *Ann. Sci. Nat. 9 ser. Bot.*, 5-6, 193–311.

—— (1934). La longévité des graines macrobiotiques transmise par Louis Mangin. *C.R. Acad. Sci., Paris*, 199, 1662–4.

BENNETT, N. & LOOMIS, W. E. (1949). Tetrazolium chloride as a test reagent for freezing injury of seed Corn. *Plant Physiol.*, 24, 162–74.

BERNAL, D. M. (1953–54). The use of 2, 3, 5-triphenyltetrazolium chloride for the determination of the viability of seeds. *Colombia, Ministeria Agric., Div. Invest., Inform. Tecnica*, 1, 79–129. (Abstr. in *Chem. Abstr.*, 49, No. 11782, 1955.)

BIBBY, K. M. & WILLIAMS, R. W. M. (1953). The effect of various over-winter storage conditions on *Nothofagus menziesii* seed. *For. Res. Note, N.Z.*, 1(7), 29–33. (Abstr. in *Biol. Abstr.*, 31, No. 8889, 1957.)

BINKLEY, A. M. (1958). A federal storage plan to safeguard seed. *The New York Times*, 5 January, p. X41.

BLACK, M. A. (1946). Effect of ceresan on the germination of stored linen-flax seed. *N.Z. J. Sci. Technol.*, A, 28, 217–8.

BLAKESLEE, A. F. & AVERY, A. G. (1934). Visible genes from aged seeds (Abstract). *Amer. Nat.*, 68, 466.

BLUM, P. H. (1954). Grain spoilage: A review. *Wallerstein Labs. Commun.*, 17(56), 35–45. (Abstr. in *Biol. Abstr.*, 28, No. 28300, 1954.)

BLUMER, J. C. (1910). The vitality of pine seeds in serotinous cones. *Torreya*, 10, 108–11.

BOSWELL, V. R., TOOLE, E. H., TOOLE, V. K. & FISHER, D. F. (1940). A study of rapid deterioration of vegetable seeds and methods for its prevention. *Tech. Bull. U.S. Dep. Agric.*, 708, 48 pp.

BRAND, C. J. & SHERMAN, W. A. (1913). Behavior of seed cotton in farm storage. *Bur. Plant Industr. Circ.*, 123, 11–20.

BRASE, K. D. (1948). Field nursery tests with newly harvested and one- or two-year stored Lovell peach seed. *Proc. Amer. Soc. Hort. Sci.*, 51, 258–62.

BRETT, C. C. & WESTON, W. A. R. D. (1941). Storage of grain treated with organomercury seed disinfectants. *J. Agric. Sci.*, 31, 500–17.

BREWER, H. E. & BUTT, J. L. (1950). Hygroscopic equilibrium and viability of naturally and artificially dried blue lupine seeds. *Plant Physiol.*, 25, 245–68.

BRISON, F. R. (1942). Influence of storage conditions upon the germination of onion seed. *Proc. Trans. Tex. Acad. Sci.*, 25(1941), 69–70. (Abstr. in *Biol. Abstr.*, 17, No. 1082, 1943.)

—— (1945). The storage of shelled pecans. *Bull. Tex. Agric. Exp. Sta.*, 667, 16 pp.

BROCQ-ROUSSEU—*see* ROUSSEU, N. BROCQ–.

BRODSKIS, B. (1949). Catalase, oxydation biologique et vitalité du grain. *Bull. Soc. Sci. Nat. Maroc.*, 35/37(1945/47), 9–12. (Abstr. in *Biol. Abstr.*, 23, No. 30230, 1949.)

BROWN, E. & GOSS, W. L. (1912). The germination of packeted vegetable seeds. *Circ. U.S. Bur. Pl. Ind.*, 101, 9 pp.

—— & ROBERT, A. L. (1943). *Alternaria* sp. on grain kernels killed by high temperature storage. *Phytopathology*, 33, 333–5.

—— TOOLE, E. H. & GOSS, W. L. (1934). Moisture content important factor in international trade in seeds. *Proc. Int. Seed Test. Ass.*, 6, 405–9.

BROWN, E. O. & PORTER, R. H. (1942). The viability and germination of seeds of *Convolvulus arvensis* L. and other perennial weeds. *Res. Bull. Ia. Agric. Exp. Sta.*, 294, 473–504.

BROWN, F. R. (1942). Influence of storage conditions on germination of onion seed. *Proc. Amer. Soc. Hort. Sci.*, 40, 501–3.

BROWN, J. C. (1954). Some observations on the reduction of 2, 3, 5-triphenyltetra-zolium chloride in plant tissue as influenced by mineral nutrition. *Plant Physiol.*, 29, 104–7.

BRÜCHER, H. (1948). Eine Schnellmethode zur Bestimmung der Keimfähigkeit von Samen. *Physiol. Plant.*, 1, 343–58.

BRUNSON, A. M. (1946). The effect of storage conditions on the longevity of seed Corn. *Rep. Dir. Ind. Agric. Exp. Sta.*, 59, 8–9.

BUNKER, H. J. (1946). Living seeds from Pharoah's tomb. *Sci. Dig.*, 20(1), 55–56.

BURGERSTEIN, A. (1895). On the duration of the vitality of the seeds of various grains. *Verhandl. Zool.-Bot. Ges. Wien*, 45, 414–21. (Abstr. in *Exp. Sta. Rec.*, 7, 777, 1896.)

BURGESS, J. L. (1938). Report on project to determine the percentage and duration of viability of different varieties of soybeans grown in North Carolina. *Proc. Ass. Off. Seed Anal. N. Amer.*, 23(1930/1931), 69.

BURLISON, W. L., van DOREN, C. A. & HACKLEMAN, J. C. (1940). Eleven years of soybean investigations. *Bull. Ill. Agric. Exp. Sta.*, 462, 167 pp.

BUSSARD, L. (1935). Contribution a l'étude des variations de la faculté germinative des semences au cours de leur conservation. *Ann. Agron.*, 5, 249–77.

BUSSE, I. (1935). Samenaufbewahrung im Vakuum. *Z. Forst- u. Jagdw.*, 67, 321–6.

CANDOLLE, A. DE (1846). Sur la durée relative de la faculté de germination, germer dans des graines appartenant à divers familles (Premier experience). *Ann. Sci. Nat. Ser.* 3, 6, 373–82.

CARRUTHERS, W. (1911). On the vitality of farm seeds. *J. R. Agric. Soc.*, 72, 168–83.

CARTER, D. G. & HOLMAN, L. E. (1952). Storing soybeans on the farm. *Ext. Circ. Ill. Coll. Agric.*, 692, 15 pp.

CARTER, E. P. & YOUNG, G. Y. (1945). Effect of moisture content, temperature, and length of storage on the development of 'sick' wheat in sealed containers. *Cereal Chem.*, 22, 418–28.

CARTLEDGE, J. L., BARTON, L. V. & BLAKESLEE, A. F. (1936). Heat and moisture as factors in the increased mutation rate from *Datura* seeds. *Proc. Amer. Phil. Soc.*, 76, 663–85.

—— & BLAKESLEE, A. F. (1934). Mutation rate increased by ageing seeds as shown by pollen abortion. *Proc. Nat. Acad. Sci., Wash.*, 20, 103–10.

—— —— (1935). Mutation rate from old *Datura* seeds. *Science*, 81, 492–3.

—— MURRAY, M. J. & BLAKESLEE, A. F. (1935). Increased mutation rate from aged *Datura* pollen. *Proc. Nat. Acad. Sci., Wash.*, 21, 597–600.

CARTTER, J. L. (1942). Equipment for maintaining controlled temperature and low humidity in a seed storage room. *J. Amer. Soc. Agron.*, 34, 1017–27.

CHAMPION, H. G. (1930). Storage of seed of Chir pine (*Pinus longifolia*). *Indian For.*, 56, 481–3.

CHILDS, J. F. L. & HRNCIAR, G. (1949). A method of maintaining viability of citrus seed in storage. *Proc. Flo. Hort. Soc.*, 61(1948), 64–69.

CHIRKOVSKIĬ, V. I. (1938). The viability of tobacco seed. *Tabak, Mosk.*, 8, 22–24. (Abstr. in *Chem. Abstr.*, 34, 3317, 1940.)

CHRISTENSEN, C. M. (1955). Grain storage studies. XVIII: Mold invasion of wheat stored for 16 months at moisture contents below 15 per cent. *Cereal Chem.* 32, 107–16.

—— (1955a). Grain storage studies. XXI: Viability and moldiness of commercial wheat in relation to the incidence of germ damage. *Cereal Chem.*, 32, 507–18.

—— (1957). Deterioration of stored grains by fungi. *Bot. Rev.*, 23, 108–34.

—— OLAFSON, J. H. & GEDDES, W. F. (1949). Grain storage studies. VIII: Relation of molds in moist stored cottonseed to increased production of carbon dioxide, fatty acids, and heat. *Cereal Chem.*, **26**, 109–28.

CHRISTIDIS, B. G. (1940). The viability of cotton seed as affected by its moisture content. *Emp. J. Exp. Agric.*, **8**, 148–58.

—— (1954). Seed vitality and other cotton characters as affected by the age of seed. *Plant Physiol.*, **29**, 124–31.

CHU, A. C., KING, G. S. & SHERMAN, G. D. (1953). *Macadamia* storage studies. *Prog. Notes Hawaii Agric. Exp. Sta.*, **90**, 1–9. (Abstr. in *Biol. Abstr.*, **29**. No. 2963, 1955).

CIESLAR, A. (1897). Versuche über Aufbewahrung von Nadelholzsamen unter luftdichten Verschlusse. *Centr. Ges. Forstw.*, **23**, 162–74.

CIFFERI, R. (1942). Una soperchieria inglese ai danni di Cosimo Ridolfi [A British supposition to the detriment of Cosimo Ridolfi]. *Atti Acad. Georgof. Firenze*, **8**(1), 70–73. (In Italian only; abstr. in *Biol. Abstr.*, **23**, No. 9540, 1949.)

ČIRKOVSKIJ, V. I. (1953). Der Einfluss des Alterns der Samen auf die Entwicklung der Pflanzen beim Tabak. *C.R. Acad. Sci., U.R.S.S.*, N.S., **92**, 439–42. (Abstr. in *Ber. Wiss. Biol.*, **90**, 78, 1954.)

CLARK, B. E. (1948). Nature and causes of abnormalities in onion seed germination. *Mem. N.Y.[Cornell] Agric. Exp. Sta.*, No. 282, 27 pp.

CLARK, C. F. (1940). Longevity of potato seed. *Amer. Potato J.*, **17**(6), 147–52. (Abstr. in *Exp. Sta. Record*, **83**, 767, 1940.)

CLAYTON, E. E. (1931). Effect of seed treatments on seed longevity. *Phytopathology*, **21**, 105–6.

CLEMENT, E. O. (1938). Germination of lily seed. *Lily Yearb. R. Hort. Soc.*, 1938, 156–7.

COILE, T. S. (1934). Influence of the moisture content of slash pine seeds on germination. *J. For.*, **32**, 468–9.

COKER, W. C. (1909). Vitality of pine seeds and the delayed opening of cones. *Amer. Nat.*, **43**, 677–81.

COLE, E. W. & MILNER, M. (1953). Colorimetric and fluorometric properties of wheat in relation to germ damage. *Cereal Chem.*, **30**, 378–91.

COLEMAN, C. H. (1953). Potato seed retains germination 13 years. *Crops & Soils*, **5**(5), 21.

COLEMAN, D. A. & FELLOWS, H. C. (1925). Hygroscopic moisture of cereal grains and flaxseed exposed to atmospheres of different relative humidity. *Cereal Chem.*, **2**, 275–87.

COLEMAN, F. B. & PEEL, A. C. (1952). Storage of seeds. *Qd. Agric. J.*, **74**, 265–76.

CONDON, M. Z., LAMBOU, M. G., VIGNES, J. L., LOE, J. B. & ALTSCHUL, A. M. (1949). Inhibitors of heating and deterioration in seeds. I: Ethylene chlorhydrin and related compounds. *Plant Physiol.*, **24**, 241–54.

COTTON, R. T. & FRANKENFELD, J. C. (1945). Protecting stored seed from insect attack. *U.S. Dep. Agric. Bur. Entomol. Plant Quarantine*, E-677, 15 pp.

—— —— BAYFIELD, E. G. & JOHNSON, J. A. (1946). Wheat viability loss due to fumigation and its relation to baking quality. *Northw. Miller*, **227**(2), Sec. 2, 4a–5a. (Abstr. in *Exp. Sta. Rec.*, **95**, 813, 1946.)

—— WALKDEN, H. H., WHITE, G. D. & WILBUR, D. A. (1953). Causes of outbreaks of stored-grain insects. *North Central Regional Publ.*, No. 35, *Bull. Kans. Agric. Exp. Sta.*, **359**, 36 pp.

COTTRELL, H. J. (1947). Tetrazolium salt as a seed germination indicator. *Nature, Lond.*, **159**, 748.

—— (1948). Tetrazolium salt as a seed germination indicator. *Ann. Appl. Biol.*, **35**, 123–31.

COUKOS, C. J. (1944). Seed dormancy and germination in some native grasses. *J. Amer. Soc. Agron.*, **36**, 337–45.

Cowgill, W. H. (1944). Studies on the germination, early seedling growth and nutrition of *Cinchona*. *Thesis (Ph.D.) Univ. Maryland*, 142 pp.

Craig, W. N. (1931). The germination of lily seeds. *Horticult., Boston*, 9, 393.

Crocioni, A. (1934). Influenza dell' età del seme sullo sviluppo della pianta. [Influence de l'âge de la graine sur le développement de la plante.] *Nuovi Ann. Agric.*, 14(2). (In Italian; abstr. in *Proc. Int. Seed Test. Ass.*, 8, 93, 1936.)

Crocker, W. (1906). Role of seed coats in delayed germination. *Bot. Gaz.*, 42, 265–91.

—— (1909). Longevity of seeds. *Bot. Gaz.*, 47, 69–72.

—— (1928). Storage, after-ripening, and germination of apple seeds. *Amer. J. Bot.*, 15, 625–6.

—— (1938). Life-span of seeds. *Bot. Rev.*, 4, 235–74.

—— (1948). *Growth of Plants; twenty years' research at Boyce Thompson Institute*. Reinhold Publishing Corp., New York, 459 pp.

—— & Barton, L. V. (1931). After-ripening, germination, and storage of certain rosaceous seeds. *Contr. Boyce Thompson Inst.*, 3, 385–404.

—— —— (1953). *Physiology of Seeds. An introduction to the experimental study of seed and germination problems*. Chronica Botanica Co., Waltham, Mass., U.S.A., 267 pp.

—— & Groves, J. F. (1915). A method of prophesying the life duration of seeds. *Proc. Nat. Acad. Sci.*, 1, 152–5.

—— & Harrington, G. T. (1918). Catalase and oxidase content of seeds in relation to their dormancy, age, vitality and respiration. *J. Agric. Res.*, 15, 137–74.

Crosier, W. F. (1930–3). Abnormal germination of wheat caused by organic mercurials. *Proc. Ass. Off. Seed Anal. N. Amer.*, 23–26, 284. (Abstr. in *Exp. Sta. Rec.*, 79, 641, 1938.)

Crossley, D. I. (1955). Viability of the seed of Lodgpole pine after 20 years in artificial storage. *For. Chron.*, 31(3), 250–3. (Abstr. in *Biol. Abstr.*, 31, No. 8899, 1957.)

Curtis, J. D. (1955). Effects of origin and storage method on the germinative capacity of ponderosa pine seed. *U.S. Dep. Agric. Intermountain Forest & Range Exp. Sta. Res. Note*, No. 26, 5 Mimeo. pp.

Cutler, G. H. (1940). Effect of 'clipping' or rubbing the oat grain on the weight and viability of the seed. *J. Amer. Soc. Agron.*, 32, 167–75.

D'Amato, F.—*see* Amato, F. D'.

Darlington, H. T. (1951). The seventy-year period for Dr. Beal's seed viability experiment. *Amer. J. Bot.*, 38, 379–81.

Darragh, W. H. (1931). The viability of sugar cane seed. *Agric. Gaz. N.S.W.*, 42, 852–4.

Darrow, G. M. & Scott, D. H. (1954). Longevity of blueberry seed in cool storage. *Proc. Amer. Soc. Hort. Sci.*, 63, 271.

Darsie, M. L., Elliott, C. & Peirce, G. J. (1914). A study of the germinating power of seeds. *Bot. Gaz.*, 58, 101–36.

Davies, P. W. A. & Winstanley, J. (1957). The use of 2, 3, 5-triphenyl-tetrazolium chloride as a measure of the seed viability of Mexican Grass, *Ixophorus unisetus* (Presl) Schlecht. *Trop. Agriculture, Trin.*, 34, 144–8.

Davis, W. C. (1931). Phenolase activity in relation to seed viability. *Plant Physiol.*, 6, 127–38.

Davis, W. E. (1925). The use of catalase as a means of determining the viability of seeds. *Proc. Ass. Off. Seed Anal. N. Amer., 18th Ann. Meeting*, pp. 33–39. (*Boyce Thompson Inst. Plant Res., Professional Paper*, 1(2), 6–12, 1926.)

de Candolle—*see* Candolle, A. de.

Decker, A. E. & Reitz, L. P. (1948). Germination tests with flax stored at different moisture and temperature levels. *Proc. Int. Seed Test. Ass.*, 14, 27–34.

DEETER, W. T. (1938). Old delphinium seeds. *Horticult., Boston*, 16, 204.

DEGEN, A. VON & PUTTEMANS, A. (1931). Sur l'influence du transport maritime sur la germination des semences. *Proc. Int. Seed Test. Ass.*, 3, 287–91.

DENT, T. V. (1947). Seed storage with particular reference to the storage of seed of Indian forest plants. *Indian For. Rec.*, 7(1), 1–134. Dehra Dun, India. (Rev. in *J. For.*, 47, 656–7, 1949.)

DEXTER, S. T., ANDERSON, A. L., PFAHLER, P. L. & BENNE, E. J. (1955). Responses of white pea beans to various humidities and temperatures of storage. *Agron. J.*, 47, 246–50.

DILLMAN, A. C. & TOOLE, E. H. (1937). Effect of age, condition, and temperature on the germination of flaxseed. *J. Amer. Soc. Agron.*, 29, 23–29.

DIMITRIEWICZ, N. (1876). Über die Methoden der Samenprüfung landwirtschaftlicher Kulturpflanzen. *Inaug.-Diss. Leipzig*, 34 pp.

DIXON, H. H. (1901). Vitality of seeds. *Nature, Lond.*, 64, 256–7.

DORE, J. (1955). Dormancy and viability of padi seed. *Malay. Agric. J.*, 38, 163–73.

DOROSHENKO, A. V. (1933). The method of staining the seeds of Umbelliferae to determine their germination capacity. *Bull. Appl. Bot. Pl.-Breed.*, Ser. A, No. 7, 185–93. (Abstr. in *Chem. Abstr.*, 28, 3760, 1934.)

—— (1937). Plasmolytic method of determining the germinating capacity of seeds. *Bull. Appl. Bot., Pl.-Breed.*, Ser. IV, 2, 119.

DORPH-PETERSEN—*see* PETERSEN, K. DORPH-.

DOTY, D. M., SMITH, G. M., ROACH, J. R. & SULLIVAN, J. T. (1945). The effect of storage on the chemical composition of some inbred and hybrid strains of sweet-corn. *Bull. Ind. Agric. Exp. Sta.*, 503, 31 pp.

DRAKE, V. C. (1948). Decline in viability of lettuce seed during laboratory storage *News Lett. Ass. Off. Seed Anal. N. Amer.*, 22(2), 31–33.

DUFFEE, F. W. (1937). Drying seed corn with electricity. *Agric. Engng., St. Joseph, Mich.*, 18, 149–51.

DUNGAN, G. H. (1940). Influence of age on the value of seed corn. *Trans. Ill. Acad. Sci.*, 33(2), 28–29.

—— & KOEHLER, B. (1944). Age of seed corn in relation to seed infection and yielding capacity. *J. Amer. Soc. Agron.*, 36, 436–43.

DUTT, B. K. & THAKURTA, A. G. (1953–5). IV: Viability of vegetable seeds in storage. *Trans. Bose Res. Inst.*, 19, 27–36.

DUVEL, J. W. T. (1904). The vitality and germination of seeds. *Bull. U.S. Bur. Pl. Ind.*, No. 58, 96 pp.

—— (1905). The storage and germination of wild rice seed. *Bull. U.S. Bur. Pl. Ind.*, No. 90, pt. 1, 16 pp.

—— (1909). The deterioration of corn in storage. *Circ. U.S. Bur. Pl. Ind.*, No. 43, 12 pp.

DYMOND, J. R. (1917). Paper packet seeds. *Proc. Ass. Off. Seed Anal. N. Amer.*, 9(1916), 30–35.

EBES, K. (1936). Die methode Gurewitschs zur Bestimmung der Keimfähigkeit ohne Keimprüfung mittels Dinitrobenzen. *Meded. Landb. Hoogesch., Wageningen*, 40, 19–21.

EGGEBRECHT, H. & BETHMANN, W. (1939). Das Selenfärbeverfahren im Vergleich zu der üblichen Keimprüfung insbesondere bei Wintergerste mit Keimruhe. *Angew. Bot.*, 21, 448–55.

EGGLESTON, H. (1949). Safe keeping for seeds. Here's how to build seed storage to preserve seed viability even in moist southern climate. *Sth. Seedsman*, 12(1), 14, 54.

EIDMANN, F. E. (1938). Eine neue biochemische Methode zur Erkennung des Aussaatwertes von Samen. *Proc. Int. Seed Test. Ass.*, 10, 203–11.

EL-SHISHINY—see SHISHINY, E. D. H. EL-.

ENGSTROM, A. (1948). Growing cottonwood from seed. *J. For.*, 46, 130–2.

ESBO, H. (1954). Vitality of unhulled and hulled timothy seeds under ordinary storing conditions in whole bags. *K. Landtbr. Akad. Handl., Stockh.*, 2-3, 123–48. (In Swedish, English summary, pp. 146–8.)

ESDORN, I. (1940). Untersuchungen an *Arnica montana* L., die Lebensdauer ihres Saatgutes und ihre Anhaubedingungen. *Dtsch. Heilpfl.*, 6, 18 pp.

EVANS, H. (1950). Results of some experiments on the preservation of cacao seed in viable condition. *Trop. Agriculture, Trin.*, 27, 48–55.

EVRARD, R. (1955). Contribution a l'étude des propriétés germinatives de faines par la methode au chlorure de 2-3-5 triphenyl-tetrazolium. *Bull. Inst. Agron. Gembloux*, 23(1), 49–54.

—— (1957). L'analyse de la qualité des semences par la méthode aux rayons. *Ann. Gembl.*, 63, 81-7.

EVREINOVA, T. N. & EROFEEV, N. G. (1956). Amylase of acorns stored under different conditions. *Věstnik Moskov. Univ. Ser. Biol. Pochvoved., Geol., Georgraf.*, 11(2), 39–43. (Abstr. in *Chem. Abstr.*, 51, 16745, 1957.)

EWART, A. J. (1896). Observations on the vitality and germination of seeds. *Trans. Biol. Soc. Liverpool*, 10, 185–93.

—— (1908). On the longevity of seeds. *Proc. Roy. Soc. Vict.*, 21, 1–210.

EXELL, A. W. (1931). The longevity of seeds. *Gdnrs.' Chron.*, III, 89, 283.

FAVILLI, R. (1950). The use of tetrazolium salts for the rapid determination of the germinative capacity of seeds. *Ann. Univ. Pisa, Fac. Agrar.*, 11, 57–84. (Abstr. in *Chem. Abstr.*, 46, 8196, 1952.)

FICK, G. L. & HIBBARD, R. P. (1925). A method for determining seed viability by electrical conductivity measurements. *Pap. Mich. Acad. Sci.*, 5, 95–103.

FIFIELD, C. C. & ROBERTSON, D. W. (1952). Milling, baking, and chemical properties of Marquis and Kanred wheat grown in Colorado and stored 19 to 27 years. *Agron. J.*, 44, 555–9.

FILUTOWICZ, A. & BEJNAR, W. (1954). The influence of the age of sugar beet seeds on their seeding value. (In Polish, with English summary.) *Roczn. Nauk rol.*, 69(3), 323–40. (Abstr. in *Biol. Abstr.*, 30, No. 14577, 1956.)

FINK, H. & SCHWEIGER, E. (1950). Ueber die Schnellmethoden zur Bestimmung der Keimfähigkeit, besonders nach dem Tetrazoliumverfahren. *Brauwissenschaft*, 3, 68–73.

FLEISCHER, F. (1851). *Beiträge zur Lehre von dem Keimen der Samen der Gewachse, insbesondere der Samen ökonomischer Pflanzen.* Gebrüdern Mäntier, Stuttgart, 159 pp.

FLEMION, F. (1936). A rapid method for determining the germinative power of peach seeds. *Contr. Boyce Thompson Inst.*, 8, 289–93.

—— (1938). A rapid method for determining the viability of dormant seeds. *Contr. Boyce Thompson Inst.*, 9, 339–51.

—— (1941). Further studies on the rapid determination of the germinative capacity of seeds. *Contr. Boyce Thompson Inst.*, 11, 455–64.

—— (1948). Reliability of the excised embryo method as a rapid test for determining the germinative capacity of dormant seeds. *Contr. Boyce Thompson Inst.*, 15, 229–41.

—— & POOLE, H. (1948). Seed viability tests with 2, 3, 5-triphenyltetrazolium chloride. *Contr. Boyce Thompson Inst.*, 15, 243–58.

FLORES, F. B. (1938). Viability of seeds of cotton as affected by moisture and age under different methods of storing. *Philipp. J. Agric.*, 9(4), 347–56.

FOY, N. R. (1934). Deterioration problems in New Zealand Chewings fescue. *N.Z. J. Agric.*, 49, 10–24.

FRANCK, W. J. (1928). Bibliography of 'germination of seed'. *Offered by the Committee for Publication and Registration to the members of the Int. Seed Test. Ass., Wageningen*, 242 Mimeo pp.

FRASER, M. T. (1916). Parallel tests of seeds by germination and by electrical response. Preliminary experiments. *Ann. Bot., Lond.*, **30**, 181–9.

FREYER, E. (1934). Additional data on the relation of moisture content to the increase of free fatty acid content of cottonseed in storage. *Oil & Soap*, **11**, 162–4, 176. (Abstr. in *Chem. Abstr.*, **28**, 6006–7, 1934.)

GADD, I. (1944). Vital colouring of pea seeds by means of malachite green. *Proc. Int. Seed Test. Ass.*, **13**(1941/3), 5–76.

—— (1953). Report of the Biochemical and Seedling Vigour Test Committee. *Proc. Int. Seed Test. Ass.*, **18**, 289–304.

—— & KJAER, A. (1940). Über die Verwendbarkeit der Selen-und Indigokarmin-methoden bei der Prüfung von Frost- und Fusariumgeschädigtem Getreide. *Proc. Int. Seed Test. Ass.*, **12**, 140–9.

GAIN, E. (1924). Anomalies des *Helianthus* issus de graines chauffées de 120° à 150°C. *C.R. Acad. Sci., Paris*, **178**, 865–7.

—— (1927). Action des températures élevées sur les graines, et morphologie des plantes issues d'embryons chauffés de 115° à 155°C. *Rev. Gén. Bot.*, **39**, 234–53, 306–29.

GANE, R. (1941). The water content of wheats as a function of temperature and humidity. *J. Soc. Chem. Ind., Lond.*, **60**, 44–46.

—— (1948). The water content of the seeds of peas, soybeans, linseed, grass, onion and carrot as a function of temperature and humidity of the atmosphere. *J. Agric. Sci.*, **38**, 81–83.

—— (1948*a*). Effect of temperature, humidity, and atmosphere on viability of Chewing's fescue grass seed in storage. *J. Agric. Sci.*, **38**, 90–92.

GARDNER, R. C. B. (1937). Storage of acorns. *Quart. J. For.*, **31**(1), 32–33. (Abstr. in *Biol. Abstr.*, **12**, No. 5383, 1938.)

GARNER, F. H. & SANDERS, H. G. (1935). Investigations in crop husbandry. II: On the age of seed beans. *J. Agric. Sci.*, **25**, 361–8.

GEORGE, E. J. (1937). Storage and dewinging of American elm seed. *J. For.*, **35**, 769–72.

GERM, H. (1943). Die Keimfähigkeit der Sojabohne. *Forschungsdienst*, **15**(1/2), 68–78. (Abstr. in *Biol. Abstr.*, **23**, No. 12599, 1949.)

—— (1957). Report of the Biochemical Viability Test Committee. *Proc. Int. Seed Test. Ass.*, **22**, 302–21.

GHOSH, T., BASAK, M. & KUNDU, B. C. (1951). Effect of insolation and chemical treatment in relation to storing of jute seeds. *Indian Phytopath.*, **4**(1), 38–44. (Abstr. in *Biol. Abstr.*, **27**, No. 4871, 1953.)

GIERSBACH, J. (1937). Some factors affecting germination and growth of gentian. *Contr. Boyce Thompson Inst.*, **9**, 91–103.

—— & CROCKER, W. (1932). Germination and storage of wild plum seeds. *Contr. Boyce Thompson Inst.*, **4**, 39–52.

GISQUET, P., HITIER, H., IZARD, C. & MOUNAT, A. (1951). Mutations naturelles observées chez *N. tabacum* L. et mutations expérimentales provoquées par l'extrait à froid de graines vielles prématurement. *Ann. Inst. Exp. Tab. Bergerac*, **1**(2), 5–36. (Abstr. in *Biol. Abstr.*, **26**, No. 9591, 1952.)

GÖKSIN, A. (1942). Altersermittlung bein Saatgut der Fichte und Kiefer. *Forstwiss. Zbl.*, **64**(5), 111–17. (Abstr. in *Biol. Abstr.*, **17**, No. 22456, 1943.)

GONZÁLEZ-RIOS, P. (1929). La germinación de la semilla de café. *Rev. Agric. P. Rico*, **22**(10), 147–8.

GOODSELL, S. F. (1948). Triphenyltetrazolium chloride for viability determination of frozen seed corn. *J. Amer. Soc. Agron.*, 40(5), 432–42.

—— HUEY, G. & ROYCE, R. (1955). The effect of moisture and temperature during storage on cold test reaction of *Zea mays* seed stored in air, carbon dioxide, or nitrogen. *Agron. J.*, 47(2), 61–64.

GORMAN, L. W. & GREENWOOD, R. M. (1951). Effect of moisture content on the viability of Italian ryegrass seed in storage. *N.Z. J. Sci. Tech.*, 33, 58–61.

GOSS, W. L. (1931). Viability of seed as shown by laboratory tests in soil. *Month. Bull. Calif. Dep. Agric.*, 20, 319–21.

—— (1937). Germination of flower seeds stored for ten years in the California State Seed Laboratory. *Bull. Calif. Dep. Agric.*, 26, 326–33.

—— (1939). Germination of buried weed seeds. *Bull. Calif. Dep. Agric.*, 28, 132–5.

—— & BROWN, E. (1939). Buried red rice seed. *J. Amer. Soc. Agron.*, 31, 633–7.

—— —— (1940). Buried red rice seed. *J. Amer. Soc. Agron.*, 32, 974.

GRACANIN, M. (1927). Über das Verhältnis zwischen der Katalase-aktivität und der Samenvitalität. *Biochem. Z.*, 180, 205–10.

GRIFFITHS, A. E. (1942). The viability of lettuce seed: a physiological and microchemical study. *Mem. N.Y. [Cornell] Agric. Exp. Sta.*, No. 245, 39 pp.

GRIMM, M. G., PREDTETSHENSKAJA, A. A., TSCHIZOVA, A. M. & EGOROVA, A. A. (1928). Über die Anwendung der 'Vitalfärbung' zur Bestimmung der Keimfähigkeit der Samen. *Act. 5th Int. Seed Test. Conf.*, Rome, p. 400–3.

GRISCH, A. & KOBLET, R. (1931). Zur Frage der Beurteilung der Keimkraft von Coniferensamen auf Grund der Katalase-bestimmung. *Proc. Int. Seed Test. Ass.*, 3(15/17), 60–74.

GUILLAUMIN, A. (1928). Le maintien des graines dans un milieu privé d'oxygène comme moyen de prolonger leur faculté germinative. *C.R. Acad. Sci.*, Paris, 187, 571–2.

GUILLEMET, R. (1931). Sur le pouvoir lipolytique de différentes variétés de graines de ricin; facteurs susceptibles de le modifier. *C.R. Soc. Biol.*, Paris, 108, 779–81.

GUNTHARDT, H., SMITH, L., HAFERKAMP, M. E. & NILAN, R. A. (1953). Studies on aged seeds. II: Relation of age of seeds to cytogenetic effects. *Agron. J.*, 45, 438–41.

GUPPY, H. B. (1897). On the postponement of germination of seeds of aquatic plants. *Proc. R. Phys. Soc., Edinb.*, 13, 344–59.

GUREWITSCH, A. (1935). Über eine Methode zur Bestimmung der Keimfähigkeit ohne Keimprafung. *Ber. Dtsch. Bot. Ges.*, 53, 303–18.

—— (1937). Über die Dinitrobenzol-Methode zur Bestimmung der Keimfähigkeit ohne Keimprüfung. II: Mitt. *Ber. Dtsch. Bot. Ges.*, 55, 54–58.

GUSTAFSON, P. R. (1951). A rapid method for determining the viability of barley. *Cereal Chem.*, 28, 343–6.

GUSTAFSSON, Å. (1937). Der Tod als ein nuklearer Prozess. *Hereditas, Lund*, 23, 1–37.

HAACK, O. (1909). Der Kiefernsamen. *Z. Forst- u. Jagdw.*, 41, 353–81.

HABER, E. S. (1950). Longevity of the seed of sweet corn inbreds and hybrids. *Proc. Amer. Soc. Hort. Sci.*, 55, 410–12.

HAFERKAMP, M. (1949). Germination and other tests on superannuated cereal seeds. *Proc. Ass. Off. Seed Anal. N. Amer.*, 39, 111–14.

HAFERKAMP, M. E., SMITH, L. & NILAN, R. A. (1953). Studies on aged seeds. I: Relation of age of seed to germination and longevity. *Agron. J.*, 45, 434–7.

HAIGH, J. C. (1940). The propagation of water hyacinth (*Eichhornia crassipes* Solms.) by seed. *Trop. Agric. Ceylon*, 94, 296–7.

—— (1952). A note on the viability of potato seeds. *Ann. Bot., Lond.*, 16, 317–19.

HAMID, M. A. (1938). Longevity of cotton seed delinted with sulphuric acid. *Emp. Cott. Gr. Rev.*, 15, 312–14. (Abstr. in *Chem. Abstr.*, 33, 5026–7, 1939.)

M   177

HAO, K.-S. (1939). Über Saatgutprüfung auf biochemischem Wege. *Z. Forst- u. Jagdw.*, 71, 141–56.

—— (1939*a*). Über Saatgutprüfung auf biochemischem Wege. *Z. Forst- u. Jagdw.*, 71, 249–69.

HARRINGTON, G. T. & CROCKER, W. (1918). Resistance of seeds to desiccation. *J. Agric. Res.*, 14, 525–32.

—— & HITE, B. C. (1923). After-ripening and germination of apple seeds. *J. Agric. Res.*, 23, 153–61.

HARRIS, R. H. & WALSTER, H. L. (1953). Observations on a sixty-four year old wheat sample. *Cereal Chem.*, 30, 58–62.

HARRISON, B. J. (1956). Seed storage. *Rep. Innes Hort. Instn.*, 46, 15–16. (Abstr. in *Biol. Abstr.*, 31, No. 25697, 1957.)

—— & McLEISH, J. (1954). Abnormalities of stored seed. *Nature, Lond.*, 173, 593–4.

HARROLD, M. E. (1935). Domestic pansy seeds. *Flor. Rev.*, 75(1935), 58–59, 66–67.

HAUT, I. C. & GARDNER, F. E. (1935). The influence of pulp disintegration upon the viability of peach seeds. *Proc. Amer. Soc. Hort. Sci.*, 32(1934), 323–7.

HEINRICH, M. (1913). Der Einfluss der Luftfeuchtigkeit, der Wärme und der Sauerstoffs der Luft auf lagerndes Saatgut. *Landw. Ver. Sta.*, 81, 289–376.

HEIT, C. E. (1943). Removing embryos of many seeds hastens germination tests. *Farm Res. [Geneva], N.Y.*, 9(3), 11, 20.

—— (1943*a*). Snow-on-the-mountain (*Euphorbia marginata*) seed data—a rapid method of detecting viability. *Proc. Ass. Off. Seed Anal. N. Amer.*, 34(1942), 78–82.

—— (1946). Fruit tree seed testing aids the nurseryman. *Farm Res. [Geneva], N.Y.* 12(4), 11–12.

—— (1947). Physiology of germination. *Rep. N.Y. St. Agric. Exp. Sta.*, 1946, 58–59.

—— (1947*a*). Physiology of germination. *Rep. N.Y. St. Agric. Exp. Sta.*, 1947, 49–50.

—— (1955). The excised embryo method for testing germination quality of dormant seed. *Proc. Ass. Off. Seed Anal. N. Amer., 45th Meeting*, pp. 108–17.

—— (1957). Laboratory germination and vigor as compared to soil tests and field performance in flower seed. *Proc. Ass. Off. Seed Anal. N. Amer.*, 47, 126–35.

—— & ELIASON, E. J. (1940). Coniferous tree seed testing and factors affecting germination and seed quality. *Tech. Bull. N.Y. St. Agric. Exp. Sta.*, 255, 45 pp.

—— & LEGNINI, C. N. (1945). Germination problems. *Rep. N.Y. St. Agric. Exp. Sta.*, 1944–5, 60–61.

—— & NELSON, C. (1941). Approximate germination tests of dormant seeds by excising embryos. *Proc. Ass. Off. Seed Anal. N. Amer.*, 33, 87–89.

HENRICI, M. (1943). On the possibility of using sodium selenite for seed testing of Karoo bushes. *S. Afr. J. Sci.*, 39, 152–4.

HIBBARD, R. P. & MILLER, E. V. (1928). Biochemical studies on seed viability. I: Measurements of conductance and reduction. *Plant Physiol.*, 3, 335–52.

—— & STREET, O. E. (1929). Biochemical studies on seed viability. II: Chemical constituents operating in reduction. *Pap. Mich. Acad. Sci.*, 9, 139–62.

HILDRETH, R. J. & SORENSON, J. W., Jr. (1957). Profits and losses from on-farm drying and storage of rice in Texas. *Bull. Tex. Agric. Exp. Sta.*, 865, 15 pp.

HILF, R. & ROHMEDER, E. (1955). Untersuchungen über das Verhalten von Tetrazoliumchlorid bei der Keimfähigkeitprüfung forstlicher Sämereien. *Forstwiss. Zbl.*, 74, 279–92.

HILTNER, L. (1891). Prüfung der Saatwasen. *Jber. AgrikChem.*, 14, 247–8.

HITCHCOCK, A. E., ZIMMERMAN, P. W., KIRKPATRICK, H. Jr., & EARLE, T. T. (1949). Water hyacinth: its growth, reproduction, and practical control by 2, 4-D. *Contr. Boyce Thompson Inst.*, 15, 363–401.

—— —— —— —— (1950). Growth and reproduction of water hyacinth and alligator weed and their control by means of 2, 4-D. *Contr. Boyce Thompson Inst.*, 16, 91-130.

HOFFPAUIR, C. L., PETTY, D. H. & GUTHRIE, J. D. (1947). Germination and free fatty acid in individual cotton seeds. *Science*, 106, 344-5.

HOLMES, A. D. (1953). Germination of seeds removed from mature and immature butternut squashes after seven months of storage. *Proc. Amer. Soc. Hort. Sci.*, 62, 433-6.

HOOVER, M. W. (1955). Influence of maturity and storage treatments upon the ascorbic acid content of the seeds of southern peas. *Food Res.*, 20(5), 469-73.

HOPKINS, E. F., RAMIREZ-SILVA, F. J., PAGAN, V. & VILLAFANE, A. G. (1947). Investigations on the storage and preservation of seed in Puerto Rico. *Bull. P.R. (Fed.) Agric. Exp. Sta.*, 72, 47 pp.

HOWE, R. W. & OXLEY, T. A. (1944). The use of carbon dioxide production as a measure of infestation of grain by insects. *Bull. Ent. Res.*, 35, 11-22.

HUELSEN, W. A. (1954). *Sweet corn.* Interscience Publishers, Inc., New York, 409 pp.

HUMMEL, B. C. W., CUENDET, L. S., CHRISTENSEN, C. M. & GEDDES, W. F. (1954). Grain storage studies. XIII: Comparative changes in respiration, viability, and chemical composition of mold-free and mold-contaminated wheat upon storage. *Cereal Chem.*, 31, 143-50.

HUSS, E. (1954). Studies of the importance of water content for the quality of conifer seed during storage. *Medd. Skogsforskninst. Stock.*, 44(7), 1-60.

HYDE, E. O. C. (1949). Methods for determining the viability of various seeds by tetrazolium staining. I: Chewing's fescue. *N.Z. J. Sci. Tech.*, 31, 13-20.

—— (1952). Methods for determining the viability of various seeds by tetrazolium staining. II: Perennial ryegrass. *N.Z. J. Sci. Tech.*, 34, 195-201.

—— (1955). Methods for determining the viability of various seeds by tetrazolium staining. III: Italian ryegrass and short-rotation ryegrass. *N.Z. J. Sci. Tech.*, 37, 36-39.

ISAAC, L. A. (1934). Cold storage prolongs the life of noble fir seed and apparently increases germinative power. *Ecology*, 15, 216-17.

—— (1935). Life of Douglas fir seed in the forest floor. *J. For.*, 33, 61-66.

ISELY, D. (1944). A study of conditions that affect the germination of *Scirpus* seeds. *Mem. N.Y. [Cornell] Agric. Exp. Sta.*, No. 257, 28 pp.

—— LOOMIS, W. E. & HUGHES, H. D. (1948). Application of chemical methods to determine seed viability. *Rep. Ia. Agric. Exp. Sta.*, 1947-48, 195.

ISSATCHENKO, B. (1931). Über die Verwendung von Farblösungen zur Untersuchung der Keimfähigkeit der Samen. *Fortschr. Landw.*, 6, 257-8.

JANIŠEVSKII, D. E. & PERVUHINA, N. V. (1941). [Prolonging the life of seeds which lose their viability quickly.] *Sovetsk. Bot.*, 1941(3), 80-86. (In Russian only; abstr. in *Biol. Abstr.*, 21, No. 25989, 1947.)

JENSEN, C. (1941). *Is it possible that seeds through treatment with light may keep their germinating power through a longer span of years than normal?* J. D. Qvist & Co., Copenhagen, 16 pp.

JENSEN, C. O., SACKS, W. & BALDAUSKI, F. A. (1951). The reduction of triphenyl-tetrazolium chloride by dehydrogenases of corn embryos. *Science*, 113, 65-66.

JOHNSON, J., MURWIN, H. F. & OGDEN, W. B. (1930). The germination of tobacco seed. *Res. Bull. Wis. Agric. Exp. Sta.*, 104, 15 pp.

JOHNSON, L. P. V. (1946). Effect of humidity on the longevity of *Populus* and *Ulmus* seeds in storage. *Can. J. Res.*, 24, 298-302.

—— (1947). Embryonic reaction to sodium biselenite as a test of seed vitality. *J. Amer. Soc. Agron.*, 39, 943-7.

JONES, D. B., DIVINE, J. P. & GERSDORFF, C. E. F. (1942). The effect of storage of corn on the chemical properties of its proteins and on its growth-promoting value. *Cereal Chem.*, 19, 819–30.

—— & GERSDORFF, C. E. F. (1939). The effect of storage on the proteins of seeds and their flours. Soybeans and wheat. *J. Biol. Chem.*, 128, xlix–l.

—— —— (1941). The effect of storage on the protein of wheat, white flour, and whole wheat flour. *Cereal Chem.*, 18, 417–34.

JONES, H. A. (1920). Physiological study of maple seeds. *Bot. Gaz.*, 69, 127–52.

JONES, J. W. (1926). Germination of rice seed as affected by temperature, fungicides, and age. *J. Amer. Soc. Agron.*, 18, 576–92.

JOSEPH, H. C. (1929). Germination and vitality of birch seeds. *Bot. Gaz.*, 87, 127–51.

—— (1929a). Germination and keeping quality of parsnip seeds under various conditions. *Bot. Gaz.*, 87, 195–210. (*Contr. Boyce Thompson Inst.*, 2, 115–30, 1929).

JOZEFOWICZ, M. (1930). Some observations on tomato plants from seeds submitted to high temperatures. *Ann. Appl. Biol.*, 17, 514–21.

JUEL, I. (1941). The auxin content of seeds of various ages, and some studies on the stability of auxin. *Planta*, 32, 227–33.

JUSTICE, O. L. (1944). Viability and dormancy in seeds of *Polygonum amphibium* L., *P. coccineum* Muhl. and *P. hydropiperoides* Michx. *Amer. J. Bot.*, 31, 369–77.

—— & WHITEHEAD, M. D. (1942). Viability of velvetbean, *Stizolobium* spp., seed as affected by date of harvest, weathering, storage, and lodging. *J. Amer. Soc. Agron.*, 34, 1000–9.

KARPER, R. E. & JONES, D. L. (1936). Longevity and viability of sorghum seed. *J. Amer. Soc. Agron.*, 28, 330–1.

KEARNS, V. & TOOLE, E. H. (1939). Temperature and other factors affecting the germination of fescue seed. *Tech. Bull. U.S. Dep. Agric.*, 638, 36 pp.

—— —— (1939a). Relation of temperature and moisture content to longevity of Chewings fescue seed. *Tech. Bull. U.S. Dep. Agric.*, 670, 27 pp.

KEVORKIAN, A. G. (1941). Cinchona production. *Rep. P.R. Agric. Exp. Sta.*, pp. 9–10.

KIDD, F. (1914). The controlling influence of carbon dioxide in the maturation, dormancy, and germination of seeds. Part II. *Proc. R. Soc.*, 87, 609–25.

KIESSELBACH, T. A. (1937). Effects of age, size, and source of seed on the corn crop. *Bull. Neb. Agric. Exp. Sta.*, 305, 16 pp.

—— (1939). Effect of artificial drying upon the germination of seed corn. *J. Amer. Soc. Agron.*, 31, 389–496.

KIKUCHI, M. (1954). Studies on the decrepitude of seeds by use of embryo transplanting method. I: On the old seeds of wheat. *Bull. Coll. Agric. Utsonomiya*, 2(2) 277–92. (In Japanese, with English summary, p. 277.)

KINCAID, R. R. (1943). Effect of storage conditions on the viability of tobacco seed. *J. Agric. Res.*, 67, 407–10.

KJAER, A. (1945). Laboratory methods of determining the germinating capacity of lupine seed compared with the field germination. *Tidsskr. Planteavl.*, 49(3), 429–44. (In Danish, with English summary, pp. 443–4.)

—— (1948). Germination of buried and dry stored seed. II: 1934–44. *Proc. Int. Seed Test. Ass.*, 14, 19–26.

—— (1950). Storage of *Larix* seeds. *Proc. Int. Seed Test. Ass.*, 16, 95–97.

KNECHT, H. (1931). Über die Beziehungen zwischen Katalaseaktivität und Vitalität in ruhenden Samen. *Bei. Bot. Zbl.*, 48, 229–313.

KNUDSON, L. (1941). Investigations on delayed photosynthesis in chlorophyll-bearing embryos of orchids, and other studies on orchid seed. *Ann. Rep. N.Y. [Cornell] Agric. Exp. Sta.*, 1940, 117.

KOEHLER, B. (1938). Effect of prolonged storage of treated seed corn (Abstract). *Phytopathology*, 28, 13.

BIBLIOGRAPHY

KOEHLER, B. & BEVER, W. M. (1956). Effect of fungicide and of storage temperature on fungicide injury to wheat seed. *Plant Dis. Reptr.*, 40, 490–2.

KONDO, M. (1926). Über die Dauer der Erhaltung der Keimkraft bei verschiedenen Samenarten in Japan. *Ber. Ohara Inst.*, 3, 127–33.

—— (1926a). Über die Erhaltung der Keimkraft von Sämereien und über Trocknungsmittel. *Ber. Ohara Inst.*, 3, 147–51.

—— & ISSHIKI, S. (1936). Storage of rice. XIV: Removal of moisture from the air in a granary and the hulled rice stored therein by a desiccating material. *Ber. Ohara Inst.*, 7, 227–37.

—— & KASAHARA, Y. (1941). [A report on storage of seeds for 10 yrs. with calcium chloride.] *Rep. Ohara Inst.*, 32, 304–10. (In Japanese; abstr. in *Biol. Abstr.*, 24, No. 10581, 1950.)

—— —— (1944). [Storage of seeds of forest trees. II: Seeds of the Japanese cedar, *Cryptomeria japonica*, and the Japanese cypress, *Chamaecyparis obtusa*. III: Further studies on seed of Cupuliferae.] *Rep. Ohara Inst.*, 36, 497–536. (In Japanese; abstr. in *Biol. Abstr.*, 24, No. 4974, 1950.)

—— & OKAMURA, T. (1931). Storage of rice. V: On the influence of a desiccating material upon the preservation of the germinating power of hulled rice having different moisture contents and stored at different temperatures. *Ber. Ohara Inst.*, 5, 221–42.

—— —— (1932/1933). Storage of rice. VI: Physical and biochemical studies of hulled rice stored in straw bags. *Ber. Ohara Inst.*, 5(3/4), 395–406.

—— —— (1932/1933a). Storage of rice. VII: On the influence of varying moisture content and germinating power upon the preservation of vitamin-B in hulled rice. *Rep. Ohara Agric. Inst.*, 5(3/4), 407–12.

—— —— (1934). Storage of rice. IX: Relation between varying moisture content and change in quality of hulled rice stored in containers air-tight as well as with carbon-dioxide. *Ber. Ohara Inst.*, 6, 149–74.

—— —— (1934a). Storage of rice. X: Studies on four lots of unhulled rice stored forty-six to eighty-four years in granaries. *Ber. Ohara Inst.*, 6, 175–85.

—— —— (1934b). Storage of rice. XI: Studies on Hoshii stored hermetically sealed for twenty-three years on a mountain. *Ber. Ohara Inst.*, 6, 331–4.

—— —— (1934c). Storage of rice. XII: Storage of rice in tin containers with calcium chloride, with special reference to the underdried product. I. *Ber. Ohara Inst.*, 6, 335–9.

—— —— (1935). Storage of rice. XIII: Storage of rice in tin containers with calcium chloride, with special reference to the underdried product. II. *Ber. Ohara Inst.*, 7, 99–102.

—— —— (1937). Storage of rice. XVI: Storage of rice in concrete silos for five years. *Ber. Ohara Inst.*, 7, 471–81.

—— —— (1938). Storage of rice. XVIII: Relation between varying moisture content and change in quality of hulled rice, with special reference to the underdried product. *Ber. Ohara Inst.*, 8, 11–34.

—— —— (1938a). Storage of rice. XX: Studies on unhulled rice stored about one hundred years in a granary. *Ber. Ohara Inst.*, 8, 47–52.

—— TAKAHASHI, R., TERASAKA, Y. & ISSHIKI, S. (1938). Storage of rice. XIX: Removal of moisture from hulled rice by heated air. *Ber. Ohara Inst.*, 8, 35–46.

—— —— —— (1941). [On the storage of seeds of forest trees. I.] *Amer. Sci.*, 32, 283–303. (In Japanese; abstr. in *Japan. J. Botan. Abstr.*, 11(4), (141)–(142), 1941.)

—— & TERASAKA, Y. (1936). Storage of rice. XV: Comparison of calcium oxide and calcium chloride as a desiccating material for rice stored in tin containers. *Ber. Ohara Inst.*, 7, 329–34.

Kopeĭkovskiĭ, V. M., Zabelina, Z. K. & Shcherbakov, V. G. (1956). Influence of moisture and some of the biochemical factors on the conditions of storage for flax seeds of high oil content. *Maslob.-Zhirov. Prom.*, 22(6), 9–12. (Abstr. in *Chem. Abstr.*, 51, 2128, 1957.)

Kornfeld, A. (1930). Untersuchung der Keimfähigkeit mit Hilfe von Farblösungen. *Fortschr. Landw.*, 5(20), 682–3.

Kreitlow, K. W. & Garber, R. J. (1946). Viability of stored seeds of forage crops treated with different fungicides. Abstr. in *Phytopathology*, 36, 403.

Kretovich, V. L. (1945). Physiologico-biochemical bases for the storage of grain. *Acad. Sci. S.S.S.R., Inst. Biochem., Moscow-Leningrad*, 136 pp. (Abstr. in *Chem. Abstr.*, 40, 1237–8, 1946.)

—— & Starodubtseva, A. I. (1956). Respiration and lipolytic activity of stored oil-bearing seeds. *Biokhim. Zerna, Sbornik*, No. 3, 179–87. (Abstr. in *Chem. Abstr.*, 51, 8914, 1957.)

Kreyer, T. (1939). [The storage of seeds of the quinine tree.] *Sovetsk. Subtrop.*, 1939(1), 49. (In Russian only.)

Krstitch, M. (1951/1952). La détermination de la vitalité des embryons de *Pinus nigra* Arn. sur le milieu à la gélatine tannisée. *Proc. Int. Seed Test. Ass.*, 17, 20–22.

Kugler, I. (1952). Keimfähigkeit und Abgabe fluoreszierender Stoffe bei Samen. *Naturwissenschaften*, 39(9), 213. (Abstr. in *Biol. Abstr.*, 27, No. 4678, 1953.)

Kuhn, R. & Jerchel, D. (1941). Über Invertseifen. VIII: Mitt. Reduktion von Tetrazolium-salzen durch Bakterien, gärende Hefe und keimende Samen. *Ber. Dtsch. Chem. Ges.*, 74, 949–52. (Abstr. in *Bot. Zbl.*, 178, 40, 1942.)

Kullen, B. (1941). Behavior of certain enzymes during the storage of wheat and its mill products. *Vorratspfl. u. Lebensmittelforsch.*, 4, 421–7. (Abstr. in *Chem. Abstr.*, 37, 2084, 1943.)

Kunzé, R. E. (1882). The germination and vitality of seeds. *Torrey Bot. Cl. Separate.*, 14 pp.

Kurata, H., Ogasawara, K. & Frampton, V. L. (1957). Microflora of milled rice. *Cereal Chem.*, 34, 47–55.

Kurokami, T., Ebihara, T. & Takematsu, T. (1947). Studies on the development of keeping quality of chestnut fruits by delaying their germination with phytohormone treatments. [In Japanese, with English title.] *J. Hort. Ass., Japan*, 16, 129–36.

Lafferty, H. A. (1931). The loss of vitality in stored farm seeds. *J. Dep. Agric. Irish Free St.*, 30, 237–45.

—— (1942). The moisture content of wheat in relation to its bushel weight and keeping quality. *J. Dep. Agric. Irish Free St.*, 39, 230–44. (Abstr. in *Chem. Abstr.*, 37, 6749–50, 1943.)

Lakon, G. (1940). Die topographische Selenmethode, ein neues Verfahren zur Feststellung der Keimfähigkeit der Getreidefrüchte ohne Keimversuch. *Proc. Int. Seed Test. Ass.*, 12, 1–18.

—— (1942). Topographischer Nachweis der Keimfähigkeit der Getreidefrüchte durch Tetrazoliumsalze. *Ber. Dtsch. Bot. Ges.*, 60, 299–305.

—— (1942a). Topographischer Nachweis der Keimfähigkeit von Mais durch Tetrazoliumsalze. *Ber. Dtsch. Bot. Ges.*, 60, 434–44.

—— (1949). The topographical tetrazolium method for determining the germinating capacity of seeds. *Plant Physiol.*, 24, 389–94.

—— (1950). Die 'Triebkraft' der Samen und ihre Feststellung nach dem topographischen Tetrazolium-Verfahren. *Saatgut-Wirtsch.*, 2(2), 37–39. (Abstr. in *Biol. Abstr.*, 25, No. 21359, 1951).

—— (1954). The application of a combined tetrazolium solution in topographical germination methods. *Saatgut-Wirtsch.*, 6, 207–8. (Abstr. in *Chem. Abstr.*, 49, 9749, 1955.)

LAKON, G. & BULAT, H. (1955). Die Feststellung der Keimfähigkeit der Kompositen-fruchte nach dem Topographischen Tetrazolium-Verfahren. *Saatgut-Wirtsch.*, 7, 201–4.

LAMBOU, M. G. (1953). 2, 3, 5-Triphenyltetrazolium chloride as a rapid indicator of viability in cotton-seed. *Science*, 117, 690–3.

—— KING, G. S. & CONDON, M. Z. (1948). Effect of chemical treatment prior to storage on viability and growth of cottonseed. *Science*, 108, 563–4.

LAPINE, L. J. & MILBERG, E. (1948). A study of the effect of moisture on the longevity of imported Chewings fescue seed under laboratory and warehouse conditions. *Proc. Ass. Off. Seed. Anal. N. Amer.*, 38, 62–65.

LARGE, J. R., FERNHOLZ, D. F., MERRILL, S., Jr. & POTTER, G. F. (1947). Longevity of tung seed as affected by storage temperatures. *Proc. Amer. Soc. Hort. Sci.*, 49, 147–50.

LARTER, L. N. H. (1947). Maize seed storage. *Trop. Agriculture, Trin.*, 1947(4/6), 40–45.

LAUGHLAND, J. & LAUGHLAND, D. H. (1939). The effect of age on the vitality of soybean seed. *Sci. Agric.*, 20, 236–7.

LEACH, W. (1943). Metabolism of cereal grains. II: Effect of age and kernel size on the course of respiration of wheat during early germination stages. *Canad. J. Res.*, 21C, 289–96.

LEGGATT, C. W. (1929–30). Catalase activity as a measure of seed viability. *Sci. Agric.*, 10, 73–110.

—— (1933). A further note on catalase activity as a measure of seed viability. *Canad. J. Res.*, 9, 571–3.

LESAGE, P. (1922). Sur la détermination de la faculté germinative autrement que par la germination des graines. *C.R. Acad. Sci., Paris*, 174, 766–7.

LEUS, F. P. (1930). Relation of age of farm crop seeds to production. *Thesis from Coll. Agric.*, No. 321. *Exp. Sta. Contrib.*, No. 682. (Abstr. in *Philipp. Agric.*, 19, 411–12, 1930.)

LEWIS, D. (1953). Chromosomal abnormalities and reduced yield from stored seed. *Rep. Innes Hort. Instn.*, No. 44, pp. 14–15.

LIBBY, W. F. (1951). Radiocarbon dates, II. *Science*, 114, 291–6.

—— (1954). Chicago radiocarbon dates, IV. *Science*, 119, 135–40.

LINDEMANN, E. (1953). The change in fat content of sorghum and maize germ during storage. *Die Stärke*, 5, 139–43. (Abstr. in *Chem. Abstr.*, 48, 1483, 1954.)

LINDSTROM, E. W. (1942). Inheritance of seed longevity in maize inbreds and hybrids. *Genetics*, 27, 154.

LIPKIN, B. IA. (1927). [Duration of viability of the seeds of different conifers.] *Ann. Weissruth. Akad. Landw. Gorki*, 5, 25–33. (In Russian, with German summary.) (Abstr. in *Biol. Abstr.*, 5, No. 8792, 1931.)

LISHKEVICH, M. I. (1953). The effect of moisture content of sunflower seeds on their keeping. *Maslob.-Zhirov. Prom.*, 18(9), 1–3. (Abstr. in *Chem. Abstr.*, 48, 820, 1954.)

LONG, T. E. & CROPSEY, M. G. (1941). Grain storage on the farm. *Bull. N. Dak. Agric. Exp. Sta.*, 302, 1–68. (Abstr. in *Biol. Abstr.*, 17, No. 9706, 1943.)

LOPEZ, M. (1938). Storage and germination of large-leaf mahogany seeds. *Philipp. J. For.*, 1, 397–410.

LUTHRA, J. C. (1936). Ancient wheat and its viability. *Curr. Sci.*, 4(7), 489–90.

LYNES, F. F. (1945). Polyploidy in sugar beets induced by storage of treated seed. *J. Amer. Soc. Agron.*, 37, 402–4.

McALISTER, D. F. (1943). The effect of maturity on the viability and longevity of the seeds of western range and pasture grasses. *J. Amer. Soc. Agron.*, 35, 442–53.

McCALLA, A. G., McCAIG, J. D. & PAUL, A. D. (1939). Effect of various conditions of storage on baking quality of flour. *Canad. J. Res. Sec. C, Bot. Sci.*, 17, 452–9. (Abstr. in *Biol. Abstr.*, 15, No. 5646, 1941.)

McClelland, T. B. (1944). Brief viability of tropical seeds. *Proc. Flo. Hort. Soc.*, **57**, 161–3.

McFarlane, V. H., Hogan, J. T. & McLemore, T. A. (1955). Effects of heat treatment on the viability of rice. *Tech. Bull. U.S. Dep. Agric.*, No. 1129, 51 pp.

Machacek, J. E. & Wallace, H. A. H. (1952). Longevity of some common fungi in cereal seed. *Canad. J. Bot.*, **30**, 164–9.

McHargue, J. S. (1920). The significance of peroxidase reaction with reference to the viability of seeds. *J. Amer. Chem. Soc.*, **42**, 612–15.

McKee, R. & Musil, A. F. (1948). Relation of temperature and moisture to longevity of seed of blue lupine, *Lupinus angustifolius*, Austrian Winter Fieldpea, *Pisum sativum arvense*, and Hairy Vetch, *Vicia villosa. J. Amer. Soc. Agron.*, **40**, 459–65.

MacLeod, A. M. (1950). Determination of germinative capacity of barley by means of tetrazolium salts. *J. Inst. Brew.*, **56**, 125–34. (Abstr. in *Chem. Abstr.*, **45**, 3914, 1951.)

—— (1952). Barley viability. *Brew. Guild J.*, **38**(453), 271–9. (Abstr. in *Biol. Abstr.*, **27**, 7441, 1953.)

—— (1952a). Enzyme activity in relation to barley viability. *Trans. Bot. Soc. Edinb.*, **36**(1), 18–33.

MacMasters, M. M. & Hilbert, G. E. (1944). Stored and damaged wheats for starch production. *Cereal Chem.*, **21**, 258–66.

McNeal, X. (1950). When to harvest rice for best milling quality and germination. *Bull. Ark. Agric. Exp. Sta.*, **504**, 41 pp.

McRostie, G. P. (1939). The thermal death point of corn from low temperatures. *Sci. Agric.*, **19**, 687–99.

Madsen, S. B. (1957). Investigation of the influence of some storage conditions on the ability of seed to retain its germination capacity. *Proc. Int. Seed Test. Ass.*, **22**, 423–46.

Manaresi, A. (1935). Sulla germinazione delle samare di olma campestre. *Italia Agric.*, **72**, 9. (In Italian, with French summary; abstr. in *Proc. Int. Seed Test. Ass.*, **8**, 92, 1936.)

Marras, F. (1955). Biochemical method for rapid determination of germinating capacity of seeds. *Studi sassaresi*, **3**, 108–17. (Abstr. in *Chem. Abstr.*, **51**, 6784, 1957.)

Marrero, J. (1942). A seed storage study of maga. *Caribbean For.*, **3**, 173–84.

—— (1943). A seed storage study of some tropical hardwoods. *Caribbean For.*, **4**, 99–106.

Martin, J. A. & Crawford, J. H. (1954). Germination of okra seed as affected by moisture content and methods of storage. *Proc. Ass. Sth. Agric. Wkrs.*, p. 120.

Martin, J. N. (1945). Changes in the germination of red clover seed in storage. *Proc. Ia. Acad. Sci.*, **51**(1944), 229–33.

Martin, S. C. (1948). Mesquite seeds remain viable after 44 years. *Ecology*, **29**, 393.

Mathur, P. B., Prasad, M. & Singh, K. Kirpal (1956). Studies in the cold storage of peanuts. *J. Sci. Food Agric.*, **7**, 354–60.

Mattson, A. M., Jensen, C. O. & Dutcher, R. A. (1947). Triphenyltetrazolium chloride as a dye for vital tissues. *Science*, **106**, 294–5.

Mazkov, F. F. (1936). Eine neue rasche Methode zum Erkennen lebender, toter und beschädigter Gewebe einer grunen Pflanze. *C.R. Acad. Sci. U.R.S.S.*, **1**(6), 265–6. (Abstr. in *Biol. Abstr.*, **11**, No. 1673, 1937.)

Miles, L. E. (1939). Effect of period and type of storage of cotton seed after treatment with organic mercury dusts. (Abstract.) *Phytopathology*, **29**, 754.

—— (1941). Effect of storage of treated cotton seed in closely woven cotton bags. (Abstract.) *Phytopathology*, **31**, 768–9.

MILNER, M. (1946). Utility of sulfa drugs for the inhibition of mold respiration in grain. *Science*, **104**, 463-4.

—— CHRISTENSEN, C. M. & GEDDES, W. F. (1947). Grain storage studies. VI: Wheat respiration in relation to moisture content, mold growth, chemical deterioration, and heating. *Cereal Chem.*, **24**, 182-99.

—— —— —— (1947*a*). Grain storage studies. VII: Influence of certain mold inhibitors on respiration of moist wheat. *Cereal Chem.*, **24**, 507-17.

—— & GEDDES, W. F. (1945). Grain storage studies. I: Influence of localized heating of soybeans on interseed air movements. *Cereal Chem.*, **22**, 477-83.

—— —— (1945*a*). Grain storage studies. II: The effect of aeration, temperature, and time on the respiration of soybeans containing excessive moisture. *Cereal Chem.*, **22**, 484-501.

—— —— (1946). Grain storage studies. III: The relation between moisture content, mold growth, and respiration of soybeans. *Cereal Chem.*, **23**, 225-47.

—— —— (1946*a*). Grain storage studies. IV: Biological and chemical factors involved in the spontaneous heating of soybeans. *Cereal Chem.*, **23**, 449-70.

—— LEE, M. R. & KATZ, R. (1952). Radiography applied to grain and seeds. *Food Technol., Champaign, Ill.*, **6**, 44-45.

—— & THOMPSON, J. B. (1954). Grain storage. Physical and chemical consequences of advanced spontaneous heating in stored soybeans. *J. Agric. & Food Chem.*, **2**, 303-9.

MIROV, N. T. (1936). A note on germination methods for coniferous species. *J. For.*, **34**, 719-23.

—— (1943). Storage and germination of California cork-oak acorns. *Res. Note U.S. For. Service*, **36**, 18 pp. (*Berkeley, Calif. For. and Range Exp. Sta.*)

—— (1944). Possible relation of linolenic acid to the longevity and germination of pine seed. *Nature, Lond.*, **154**, 218-19.

—— (1946). Viability of pine seed after prolonged cold storage. *J. For.*, **44**, 193-5.

MONSELISE, S. P. (1953). Viability tests with citrus seeds. *Palest. J. Bot., Rehovot*, **8**, 152-7. (Abstr. in *Biol. Abstr.*, **29**, No. 9569, 1955.)

MOORE, R. P. & SMITH, E. (1957). Tetrazolium testing. General technic. *News Lett. Ass. Off. Seed Anal. N. Amer.*, **31**(2), 33-38.

MOORJANI, M. N. & BHATIA, D. S. (1954). Storage effects on the proteins of groundnuts. *J. Sci. Industr. Res.*, **13**B, 113-14; cf. Pickett, C. A. **43**, 329*a*. (Abstr. in *Chem. Abstr.*, **48**, 12339, 1954.)

MORRISON, B. Y. (1944). New quinine from this hemisphere. *Nat. Hort. Mag.*, **23**, 21-31.

MOSS, E. H. (1938). Longevity of seed and establishment of seedlings in species of *Populus. Bot. Gaz.*, **99**, 529-42.

MUENSCHER, W. C. (1936). The germination of seeds of *Potamogeton. Ann. Bot., Lond.*, **59**, 805-21.

—— (1936*a*). Storage and germination of seeds of aquatic plants. *Bull. Cornell Agric. Exp. Sta.*, **652**, 17 pp.

MUKHERJI, D. K. (1956). Seed testing. X: Investigations on tetrazolium salt as a rapid germination tester for seeds of agricultural and vegetable crops. *Sci. & Cult.*, **21**, 390-1. (Abstr. in *Biol. Asbtr.*, **31**, No. 32462, 1957.)

MÜLLER, F. VON (1880). Vitality of *Eucalyptus* seeds. *Gdnrs.' Chron.*, N.S., **13**, 811.

MÜLLER-OLSEN, C., SIMAK, M. & GUSTAFSSON, Å. (1956). Germination analyses by the X-ray method: *Picea abies* (L.) Karst. *Medd. Statens Skogsforsöksanst. Stockh.*, **46**(1), 1-12.

MUNN, M. T., HOEFLE, O. M. & WOODBRIDGE, M. E. (1929). The quality of packet vegetable seed on sale in New York in 1926, 1927, and 1928. *Bull. N.Y. St. Agric. Exp. Sta.*, **565**, 47 pp.

MUNN, M. T. & MUNN, R. E. (1936). The quality of flower seeds on sale in New York. *Bull. N.Y. St. Agric. Exp. Sta.*, 663, 12 pp.

MYERS, A. (1935). The viability of parsnip seed. *Agric. Gaz. N.S.W.*, 46, 672.

—— (1939). Prickly pear seed may remain viable for years. Germination tests with *Opuntia* species. *Agric. Gaz. N.S.W.*, 50, 660.

—— (1940). Longevity of seed of native grasses. *Agric. Gaz. N.S.W.*, 51, 405.

—— (1942). Cold storage of onion seed. *Agric. Gaz. N.S.W.*, 53, 528.

—— (1943). Germination of stored onion seed. *Agric. Gaz. N.S.W.*, 54, 398.

—— (1946). Germination tests of stored tomato seed. *Agric. Gaz. N.S.W.*, 57, 290.

—— (1952). Germination of stored lucerne seed; results of four-year experiment. *Agric. Gaz. N.S.W.*, 63, 284–5.

—— (1954). Viability of strawberry seed. *Agric. Gaz. N.S.W.*, 65, 31.

NÁDVORNÍK, J. (1946). Použití vitálního barvení ke zkoušení semen ovocných dřevin. [Application of vital staining of seeds of fruit trees and shrubs.] *Bull. Czechosl. Acad. Agric.*, 20(3/4), 160–4. (In Czech, with Russian summary; abstr. in *Biol. Abstr.*, 21, No. 7203, 1947.)

NAKAJIMA, Y. (1921). Length of life of the seed of the genus *Salix*. *Botan. Mag.*, *Tokyo*, 35, (17)–(42). (In Japanese, with English title; abstr. in *Botan. Abstr.*, 14, 3007, 1925.)

—— (1925). Über die Keimfähigkeitsdauer der Reiskörner. *Botan. Mag.*, *Tokyo*, 39, 265. (In Japanese; resumé in German.)

—— (1926). Weitere Untersuchungen über die Lebensdauer der Weidensamen. *Sci. Repts. Tôhoku Imp. Univ.*, 4th ser., 1, 261–75.

—— (1927). Untersuchungen über die Keimfähigkeitsdauer der Samen. *Bot. Mag.*, *Tokyo*, 41, 604–32. (In Japanese; German resumé, pp. 630–32.)

NAKATOMI, S. (1936). [Relations of oxidase contents to the germination capacity of cotton seeds.] *Proc. Crop Sci. Soc. Japan*, 8, 341–8. (In Japanese; English abstr. in *Japan J. Bot.*, 9(1), 16, 1937.)

NANDA, K. K. (1950). Catalase ratio as a rapid method for determining the germinating capacity of seeds. *Curr. Sci.*, 19, 22–24.

NAVASHIN, M. (1933). Origin of spontaneous mutations. *Nature, Lond.*, 131, 436.

—— GERASSIMOVA, H. & BELAJEVA, G. M. (1940). On the course of the process of mutation in the cells of the dormant embryo within the seed. *C.R. Acad. Sci. U.R.S.S.*, 26, 948–51. (Abstr. in *Gartenbauwiss.*, 15(4), S82–83.)

—— & SHKVARNIKOV, P. (1933). Process of mutation in resting seeds accelerated by increased temperature. *Nature, Lond.*, 132, 482.

NAZAROVA, N. S. (1937). [Catalase activity as an index of the germinating capacity of seeds.] *Bull. Appl. Bot. Pl. Breed.*, Ser. IV, 2, 103–11. (In Russian, with English summary, p. 111.)

NELJUBOW, D. (1925). Über die Methoden der Bestimmung der Keimfähigkeit ohne Keimprüfung. *Ann. Ess. Semences, Leningrad*, 4(7), 31–35.

NELSON, M. L. (1940). Successful storage of southern pine seed for seven years. *J. For.*, 38, 443–4.

NELSON, O. E., Jr. & BURR, H. S. (1946). Growth correlates of electromotive forces in maize seeds. *Proc. Natl. Acad. Sci.*, 32(4), 73–84.

NEMEČ, A. & DUCHOŇ, F. (1922). Sur une méthode indicatrice permettant d'évaluer la vitalité des semences par voie biochimique. *C.R. Acad. Sci., Paris*, 174, 632–4.

—— —— (1923). Sur une nouvelle méthode biochimique pour la détermination de la force vitale des semences. *Ann. Sci. Agron., Paris*, 40, 121–50.

NICHOLSON, C. (1932). Longevity of wheat seeds. *Gdnrs.' Chron.*, 92(2399), 453.

NIETHAMMER, A. (1931). Kann die Keimfähigkeit unserer Samen ohne Keimprobe bestimmt werden? *Fortschr. Landw.*, 6(8), 258–9.

—— (1931a). Critical comparative sugar and catalase determinations on seed materials of different vitalities. *Z. PflErnähr. Düng.*, 21A, 69–86. (*Chem. Abstr.*, 25, 5688, 1931).

—— (1942). Plasmolysestudien an gärtnerisch wichtgem Saatgut. *Gartenbauwiss.*, 17, 91–94. (*Biol. Abstr.*, 18, No. 11860, 1944.)

NILAN, R. A. & GUNTHARDT, H. M. (1956). Studies on aged seeds. III: Sensitivity of aged wheat seeds to X-radiations. *Caryologia*, 8, 316–21. (*Biol. Abstr.*, 31, No. 19457, 1957.)

NILES, J. J. (1954). I: The use of triphenyl tetrazolium bromide in viability tests of rice seed. *Trop. Agriculturist*, 110, 13–20.

—— (1955). II: The use of triphenyl tetrazolium bromide in viability tests of rice seed. *Trop. Agriculturist*, 111, 279–85.

NORD, E. C. (1956). Quick testing bitterbrush seed viability. *J. Range Management*, 9(4), 193–4. (*Biol. Abstr.*, 31, No. 2349, 1957.)

NORTH, C. (1948). Artificial drying of vegetable and herbage seeds. *J. Minist. Agric.*, 54, 462–6.

NUCCORINI, R. (1930). Viability and catalases of seed. *Boll. Ist. Agr. Pisa*, 6, 319–35. (*Biol. Abstr.*, 6, No. 2213, 1932.)

NUTILE, G. E. (1957). Effect of relative humidity on the viability of lettuce and onion seeds during nine years of storage. *News Lett. Ass. Off. Seed Anal. N. Amer.*, 31(2), 11–14.

OATHOUT, C. H. (1928). The vitality of soybean seed as affected by storage conditions and mechanical injury. *J. Amer. Soc. Agron.*, 20, 837–55.

OHGA, I. (1923). On the longevity of the fruit of *Nelumbo nucifera*. *Bot. Mag.*, Tokyo, 37, 87–95.

—— (1926). On the structure of some ancient, but still viable fruits of Indian Lotus, with special reference to their prolonged dormancy. *Jap. J. Bot.*, 3, 1–20

—— (1926a). The germination of century-old and recently harvested Indian lotus fruits, with special reference to the effect of oxygen supply. *Amer. J. Bot.*, 13, 754–9. (*Contr. Boyce Thompson Inst.*, 1, 289–94, 1926.)

—— (1926b). A comparison of the life activity of century-old and recently harvested Indian lotus fruits. *Amer. J. Bot.*, 13, 766–72. (*Contr. Boyce Thompson Inst.*, 1, 295–300, 1926.)

OLAFSON, J. H., CHRISTENSEN, C. M. & GEDDES, W. F. (1954). Grain storage studies. XV: Influence of moisture content, commercial grade, and maturity on the respiration and chemical deterioration of corn. *Cereal Chem.*, 31, 333–40.

OLSEN, MÜLLER- —*see* MÜLLER-OLSEN, C.

OPPERMANN, O. (1913). Winter storage of acorns. *Forstl. Forsøksv. Danm.*, 4(2), 127–33. (Abstr. in *Exp. Sta. Rec.*, 29, 343, 1913.)

OWEN, E. B. (1956). *The storage of seeds for maintenance of viability*. Commonwealth Agricultural Bureaux, Farnham Royal, Bucks, pp. 81. (Rev. in *S. Afr. J. Sci.*, 53, 274–5, 1957.)

OXLEY, T. A. (1950). *Grain storage in East and Central Africa. Report of a survey.* (*Oct. 1948 to Jan. 1949.*) Colon. Res. Publ., No. 5, iv+43, 20 pls., London, H.M.S.O. (Abstr. in *Rev. Appl. Entomol.*, A, 40, Part 9, 260–1, 1952.)

—— (1950a). The storage and drying of cereal seeds. *J. Nat. Inst. Agric. Bot.*, 5, 465–82.

—— (1955). Grain storage in tropical climates. *World Crops*, 7, 473–7.

—— & HYDE, M. E. (1955). Recent experiments on hermetic storage of wheat. *Proc. Third Int. Bread Congr.*, Hamburg, 1955, 179–82.

PAINTER, E. P. & NESBITT, L. L. (1943). The stability of linseed oil during storage of flaxseed. *Bi-m. Bull. N. Dak. Agric. Exp. Sta.*, 5(6), 36–40.

PANIKKAR, M. R. (1947). Storage of rice. *Sci. & Cult.*, 13(2), 59–66. (*Biol. Abstr.*, 22, No. 25121, 1948.)

PAPAVIZAS, G. C. & CHRISTENSEN, C. M. (1957). Grain storage studies. XXV: Effect of invasion by storage fungi upon germination of wheat seed and upon development of sick wheat. *Cereal Chem.*, 34, 350–9.

PARKER, J. (1953). New methods for the determination of forest tree seed germinability. *J. For.*, 51, 34.

—— (1955). Tetrazolium salts as viability indicators: radiation-catalysed reactions. *Nature, Lond.*, 176, 647–8.

PARKINSON, A. H. (1948). Seed age and storage affect cucumber growth. *Fm. Res.*, 14(3), 7.

PATON, R. R. (1945). Storage of tuliptree seed. *J. For.*, 43, 764–5.

PATRICK, S. R. (1936). Some observations on the decline in viability of Chewing's fescue seed (*Festuca rubra commutata*). *Proc. Ass. Off. Seed Anal. N. Amer.*, 28(1936), 76–79.

PEARCE, J. A. (1943). Effect of storage on thiamin content and on development of rancidity in wheat germ. *Canad. J. Res.*, C., 21, 57–65.

PEDRO, A. V. SAN (1936). Influence of temperature and moisture on the viability of some vegetable seeds. *Philipp. Agric.*, 24, 649–58.

PERCIVAL, J. (1921). *The wheat plant.* Duckworth & Co., London, 463 pp.

PETERSEN, K. DORPH- (1925). Germination tests in the laboratory and in soil of cereal seed which is not 'germinating-ripe'. *Rep. Int. Seed Test. 4th Congr., Cambridge,* 7–12 July, 1924, pp. 76–82.

—— (1928). Comment les plus importantes des espèces de semences cultivées gardent-elles leur faculté germinative dans les magasins de semences ordinaires? *Proc. Int. Seed Test. Ass.,* 1, 57–71.

—— (1928a). Combien de temps les semences de *Tussilago fárfara* gardent-elles leur faculté germinative sous de différentes conditions de température? *Proc. Int. Seed Test. Ass.,* 1, 72–76.

PETERSOHN, E. (1926). Weitere Versuche über die Heranziehung der Katalasenwirkung von Samenkörnern zwecks Beurteilung der Keimfähigkeit und der Ursachen des Verlustes derselben. *Z. PflErnähr. Düng.*, 6(5), 291–4.

PETO, F. (1933). The effect of ageing and heat on the chromosomal mutation rates in maize and barley. *Canad. J. Res.*, 9, 261–4.

PHILLIS, E. & MASON, T. G. (1945). The effect of extreme desiccation on the viability of cotton seeds. *Ann. Bot., Lond.*, 9, 353–9.

PINGALE, S. V. & BALU, V. (1955). Grain storage practices in India. *Trop. Agriculture, Trin.*, 32(2), 88–94.

PLAUT, M. & GABRIELIT-GELMOND, CH. (1949). Determination of the viability of seeds of some vegetable and field crops by means of sodium selenite. *Proc. Int. Seed Test. Ass.,* 14(1948), 190–211.

—— & HALFON, A. (1954). The viability test of pea, bean, and cucumber seeds by resazurin staining. *Proc. Int. Seed Test. Ass.,* 19, 14–23.

—— —— HELLER-COHEN, O., COHEN, A. & GORDIN, A. (1957). Determination of the viability of seeds by resazurin. *Proc. Int. Seed Test. Ass.,* 22, 343–7.

—— & HELLER-COHEN, O. (1956). New methods for testing viability of seeds. *World Crops*, 8, 311, 319.

PONS, W. S., Jr., MURRAY, M. D., O'CONNOR, R. T. & GUTHRIE, J. D. (1948). Storage of cottonseed and peanuts under conditions which minimize spectrophotometric changes in the extracted oil. *J. Amer. Oil Chem. Soc.*, 25, 308–13. (*Chem. Abstr.*, 42, 8497, 1948.)

POPTSOV, A. V. (1929). Studies of the seeds of kenaph (*Hibiscus cannabis* L.). *Arb. Wiss. Forsch Inst. Kuban. Landw. Inst.*, 8, 83–125. (*Biol. Abstr.*, 6, No. 24710, 1932.)

POPTZOFF, A. (1933). [Studies in storage of tobacco seed. Part 2. The influence of air humidity, temperature, and air-tight keeping on tobacco seeds during their storage.] *U.S.S.R. Bull. Inst. Tob. Invest. Krasnodar*, **93**, 1–58. (In Russian, English summary, pp. 56–58.)

PORTER, R. H., DURRELL, M. & ROMM, H. J. (1947). The use of 2, 3, 5-triphenyl-tetrazolium chloride as a measure of seed germinability. *Plant Physiol.*, **22**, 149–59.

PRITCHARD, E. W. (1933). How long do seeds retain their germinating power? Longevity experiments in the department of agriculture. *J. Dep. Agric. S. Aust.*, **36**, 645–6.

PUNTAMBEKAR, S. V. (1947). Effect of storage on the alkaloidal content of *Strychnos nux-vomica* seeds. *Curr. Sci.*, **16**(11), 346.

QVAM, O. (1906). Zur Atmung des Getreides. Eine Relation zwischen Keimfähigkeit und Atmungsintensität. *Jber. Ver. Angew. Bot.*, **4**, 70–87.

RAMSTAD, P. E. (1942). *A study of the respiration and storage behavior of soybeans.* Thesis, Univ. Minn., 106 pp., May.

—— & GEDDES, W. F. (1942). The respiration and storage behavior of soybeans. *Tech. Bull. Minn. Agric. Exp. Sta.*, **156**, 54. (Abstr. in *Exp. Sta. Rec.*, **88**(4), 431, 1943.)

RAO, M. N., VISWANATHA, T., MATHUR, P. B., SWAMINATHAN, M. & SUBRAHMANYAN, V. (1954). Effect of storage on the chemical composition of husked, undermilled and milled rice. *J. Sci. Fd. Agric.*, **5**, 405–9.

REES, B. (1911). Longevity of seeds and structure and nature of seed coats. *Proc. Roy. Soc. Vict.*, **23**, 393–414.

REHACKOVA, O. (1954). [Yearly variations in germination percent and rapidity of germination of pine and larch seeds.] *Schweiz. Z. Forstw.*, **105**(8), 459–67. (In German, with English title; Abstr. in *Biol. Abstr.*, **29**, No. 14714, 1955.)

REYNOLDS, E. B., CORY, V. L. & DAMERON, W. H. (1938). Germination and longevity of bitterweed seed. *50th Ann. Rep. Tex. Agric. Exp. Sta.*, 1937, 82.

RICHARDS, A. V. (1952). Viability of citrus seeds. *Trop. Agriculturist*, **108**, 186–8.

RIDDELL, J. A. & GRIES, G. A. (1956). The influence of age and maturation temperature of wheat grains on plant development. *Proc. Ind. Acad. Sci.*, **66**, 62.

RITVANEN, T. (1953). The application of tetrazolium chloride to the rapid determination of the germinability of timothy seed. *Maataloust. Aikakausk.*, **25**(4), 153–9. (*Chem. Abstr.*, **48**, 9474, 1954.)

ROBBINS, W. A. & PORTER, R. H. (1946). Germinability of sorghum and soybean seed exposed to low temperature. *J. Amer. Soc. Agron.*, **38**, 905–13.

ROBERTS, L. W. (1951). Survey of factors responsible for reduction of 2, 3, 5-triphenyl-tetrazolium chloride in plant meristems. *Science*, **113**, 692–3.

ROBERTSON, D. W. & LUTE, A. M. (1933). Germination of the seed of farm crops in Colorado after storage for various periods of years. *J. Agric. Res.*, **46**, 455–62.

—— —— (1937). Germination of seed of farm crops in Colorado after storage for various periods of years. *J. Amer. Soc. Agron.*, **29**, 822–34.

—— —— & GARDNER, R. (1939). Effect of relative humidity on viability, moisture content, and respiration of wheat, oats, and barley seed in storage. *J. Agric. Res.*, **59**, 281–91.

—— —— & KROEGER, H. (1943). Germination of 20-year-old wheat, oats, barley, corn, rye, sorghum and soybeans. *J. Amer. Soc. Agron.*, **35**, 786–95.

ROBERTSON, F. R. & CAMPBELL, J. G. (1933). The increase of free fatty acid in cotton-seed. *Oil & Soap*, **10**, 146–7. (*Chem. Abstr.*, **27**, 4704, 1933.)

RODRIGO, P. A. (1935). Longevity of some farm crop seeds. *Philipp. J. Agric.*, **6**, 343–57.

—— (1939). Study on the vitality of old and new seeds of mungo (*Phaseolus aureus* Roxb.). *Philipp. J. Agric.*, **10**, 285–91.

—— & TECSON, A. L. (1940). Storing some vegetable seeds. *Philipp. J. Agric.*, **11**, 383–95.

ROE, E. I. (1948). Viability of white pine seed after 10 years of storage. *J. For.*, **46**, 900–2.

ROSE, D. H. (1915). A study of delayed germination in economic seeds. *Bot. Gaz.*, **59**, 425–44.

ROUSSEU, N. BROCQ- & GAIN, E. (1909). Oxydases et peroxydiástases des graines. *Rev. Gén. Botan.*, **20**, 55–62.

RUDOLF, P. O. (1937). Delayed germination in American elm. *J. For.*, **35**, 876–7.

RUGE, U. (1952). Über die Steigerung der Keimfähigkeit alten Saatgutes mit Hilfe von Äthylen-Chlorhydrin. *Angew. Bot.*, **26**(3/4), 162–5.

RUSCA, R. A. & GERDES, F. L. (1942). Effects of artificially drying seed cotton on certain quality elements of cotton seed in storage. *Circ. U.S. Dep. Agric.*, No. 651, 19 pp.

RUSSELL, R. C. (1956). Longevity studies with wheat seed and certain seed-borne fungi (Abstr.). *Proc. Canad. Phytopath. Soc.*, **24**, 22.

SAHS, W. W. (1957). Nebraska foundation seed. *Crops & Soils*, **9**(9), 22–23.

SAKATA, T. (1933). The use of chemicals to determine the vitality of seeds. *Seed World*, **33**(13), 64.

SAMEK, J. (1894). Duration of the vitality of some agricultural seeds. *Tirol. Landw. Bl.*, **13**(18), 161–2. (Abstr. in *Exp. Sta. Rec.*, **6**, 429–30, 1894/1895.)

SAMPIETRO, G. (1931). Prolonging the longevity of rice seed. *G. Risic.*, **21**, 1–5. (*Chem. Abstr.*, **25**, 2172, 1931.)

SARAN, A. B. (1945). On the viability of paddy seeds, *Oryza sativa. Curr. Sci.*, **14**, 271.

SARGENT, C. S. (1880). Vitality of the seeds of *Pinus contorta. Bot. Gaz.*, **5**, 54.

SAYRE, J. D. (1940). Storage tests with seed corn. *Ohio J. Sci.*, **40**, 181–5.

—— (1947). Storage tests with seed corn. *Bi-m. Bull. Farm & Home Res.*, **32**(247), 149–54. (*Biol. Abstr.*, **23**, No. 8672, 1949.)

SCHENK, R. U., MacMASTERS, M. M. & SENTI, F. R. (1957). A note on improved interpretation of the 2, 3, 5-triphenyltetrazolium chloride color test for viability as an indication of the processing value of corn. *Cereal Chem.*, **34**, 69–70.

SCHJELDERUP-EBBE, T. (1936). Uber die Lebensfähigkeit alter Samen. *Skr. Norske VidenskAkad.*, 1935, 1, 178. (*Biol. Abstr.*, **11**, No. 10520, 1937.)

SCHLOESING, A. TH. & LEROUX, D. (1943). Essai de conservation de graines en l'absence d'humidité d'air et de lumière. *C.R. Acad. Agric. Fr.*, **29**, 204–6. (Abstr. in *Ann. Agron.*, **13**, 333, 1943.)

SCHMIDT, E. W. (1934). Über das Altern des Zuckerrübensamens [On the increase of age in sugar beet seed.] *Dtsch. Zuckerindstr.*, **59**(48), 961–3. (Rev. in *Proc. Int. Seed Test. Ass.*, **7**, 86–87, 1935.)

SCHMIDT, J. L. (1955). Wheat storage research at Hutchinson, Kans., and Jamestown, N. Dak. *Tech. Bull. U.S. Dep. Agric.*, No. 1113, 98 pp.

SCHMIDT, W. (1929). Weitere Katalaseuntersuchungen als Prüfmassstab des Samen-zustands. *Z. Forst- u. Jagdw.*, **61**, 413–28.

SCHUBERT, G. H. (1954). Viability of various coniferous seeds after cold storage. *J. For.*, **52**, 446–7.

—— (1955). Effect of storage temperature on viability of sugar, Jeffrey, and ponderosa pine seed. *For. Res. Note, Calif.*, **100**, 3 pp.

SCHWASS, R. H. (1951). Assessing the value of low-germinating seed. *N.Z. J. Agric.*, **82**, 225.

SCHWEMMLE, J. (1940). Keimversuche mit alten Samen. *Z. Bot.*, **36**, 225–61.

SEMENIUK, G. & GILMAN, J. C. (1944). Relation of molds to the deterioration of corn in storage, a review. *Proc. Ia. Acad. Sci.*, **51**, 265–80.

—— NAGEL, C. M. & GILMAN, J. C. (1947). Observations on mold development and on deterioration in stored yellow dent shelled corn. *Res. Bull. Ia. Agric. Exp. Sta.*, No. 349, 253–84.

SERENKOV, G. P. & KUZNETSOVA, V. S. (1952). The change in the carbohydrate complex of acorns during storage. *Vestn. Moskov. Univ.* 7, No. 2 Ser. *Fiz.-Mat.* —*Estestven. Nauk.*, No. 1, 119–25. (*Chem. Abstr.*, 46, 9669, 1952.)

SHAMEL, A. D. & COBEY, W. W. (1907). Tobacco breeding. *Bull. U.S. Bur. Pl. Ind.*, 96, 71 pp.

SHEAR, C. B. & CRANE, H. L. (1943). Germination of the nuts of the Tung tree as affected by penetrants, substrata, depth of planting and storage conditions. *Bot. Gaz.*, 105, 251–6.

SHEDD, C. K. & WALKDEN, H. H. (1947). Grain sorghum storage. *Circ. U.S. Dep. Agric.*, No. 760, 1–48.

SHEFER-SAFONOVA, E. IĀ., KALASHĪNKOVA, M. I. & KOSTROMĪNA, A. S. (1934). Determination of the viability of seeds of trees by a staining method. *Bot. Zh. S.S.S.R.*, 19, 566–94. (*Biol. Abstr.*, 11, No. 14598, 1937.)

SHIRASAWA, H. & KOYAMA, M. (1915). Experiments on the preservation of principal forest tree seeds. *Bull. For. Exp. Sta.*, *Tokyo-Fu*, 1915, 1–13.

SHISHINY, E. D. H. EL- (1953). Effect of temperature and desiccation during storage on germination and keeping quality of *Kochia indica* seeds. *J. Expt. Bot.*, 4, 403–6.

SHKVARNIKOV, P. K. (1937). Über Erhöhung der Mutationsrate bei Weizen nach langer Aufbewahrung der Samen. *Genetica*, 19, 188–99. (Rev. in *Z. Bot.*, 32, 540–1, 1938.)

SHKVARNIKOV, P. & NAVASHIN, M. (1935). Acceleration of the process of mutation in resting seeds under the influence of increased temperature. *Biol. Zh.*, 4, 25–38.

SHUEL, R. W. (1948). Seed germinability tests with 2, 3, 5-triphenyltetrazolium chloride. *Sci. Agric.*, 28, 34–38.

SIFTON, H. B. (1920). Longevity of the seeds of cereals, clovers and timothy. *Amer. J. Bot.*, 7, 243–51.

SIMAK, M. (1957). The X-ray contrast method for seed testing. Scots Pine—*Pinus silvestris. Med. Stat. Skogsforskninst., Stockh.*, 47(4), 1–22.

—— GUSTAFSSON, Å. & GRANSTRÖM, G. (1957). Die Röntgendiagnose in der Samenkontrolle. *Proc. Int. Seed Test. Ass.*, 22, 330–43.

SIMPSON, D. M. (1935). Relation of moisture content and method of storage to deterioration of stored cottonseed. *J. Agric. Res.*, 50, 449–56.

—— (1940). Storage tests with cottonseed. (Abstract.). *Phytopathology*, 30, 707.

—— (1942). Factors affecting the longevity of cottonseed. *J. Agric. Res.*, 64, 407–19.

—— (1946). The longevity of cottonseed as affected by climate and seed treatments. *J. Amer. Soc. Agron.*, 38, 32–45.

—— (1953). Cottonseed storage in various gases under controlled temperature and moisture. *Bull. Univ. Tenn. Agric. Exp. Sta.*, No. 228, 16 pp.

—— (1953a). Longevity of cottonseed. *Agron. J.*, 45, 391.

—— & MILLER, P. R. (1944). The relation of atmospheric humidity to moisture in cottonseed. *J. Amer. Soc. Agron.*, 36, 957–9.

—— & STONE, B. M. (1935). Viability of cottonseed as affected by field conditions. *J. Agric. Res.*, 50, 435–47.

SINGH, B. N. (1941). The relationship of catalase ratio to germination of X-rayed seed as an example of pretreatments. *J. Amer. Soc. Agron.*, 33, 1014–16.

—— MATHUR, P. B. & MEHTA, M. L. (1938). Determination of catalase ratio as a rapid method of seed testing. *Trop. Agriculture, Trin.*, 15, 260–1.

SINGH, S. L. & SINGH, S. G. (1935). A study of the viability of some common winter vegetable seeds. *Agric. Live-Stk.*, *India*, 5, 670–5.

SINGLEY, M. E., KRUEGER, W. C. & MERRILL, L. G., Jr., (1954). Safe farm storage for grain and shelled corn. *Circ. N.J. Agric. Exp. Sta.*, No. 562, 11 pp.

SLUSANSCHI, H. (1939). The changes in the chemical character of grains during storage. *Anal. Inst. Cerc. Agron. Român.*, 11, 320–6. (Abstr. in *Chem. Abstr.*, 37, 4482, 1943.)

SMITH, A. J. M. & GANE, R. (1939). The water relations of seeds. *Rep. Fd. Invest. Bd.*, *Lond.*, 1938, 231–40.

SMITH, C. L., THOR, C. J. B. & ROMBERG, L. D. (1933). Effect of storage conditions on the germination of seed pecans. *Proc. Tex. Pecan Gr. Ass.*, 13, 68–71. (Abstr. in *Exp. Sta. Rec.*, 70, 782, 1934.)

SMITH, F. G. (1952). The mechanism of the tetrazolium reaction in corn embryos. *Plant Physiol.*, 27, 445–56.

SMITH, L. (1948). The effect of chaff of cereals on germination of seeds and on the growth of mold. *J. Amer. Soc. Agron.*, 40, 32–44.

SNYDER, G. B. & TUTTLE, A. P. (1937). Packet seed studies. *Bull. Mass. Agric. Exp. Sta.*, 339(1936), 74.

SOLOVJEVA, M. A. (1950). Fruit seed storage. *Sad i Ogorod.*, pp. 24–28. (Abstr. in *Brit. Abstr.*, B III, 113, 1953.)

SONAVNE, K. M. (1934). Longevity of crop seeds, Part II. *Agric. and Live-Stk.*, *India*, 4, 287–92.

SORGER-DOMENIGG, H., CUENDET, L. S., CHRISTENSEN, C. M. & GEDDES, W. F. (1955). Grain storage studies. XVII: Effect of mold growth during temporary exposure of wheat to high moisture contents upon the development of germ damage and other indices of deterioration during subsequent storage. *Cereal Chem.*, 32, 270–85.

—— & GEDDES, W. F. (1955a). Grain storage studies. XX: Relation between viability, fat acidity, germ damage, fluorescence value, and formazan value of commercial wheat samples. *Cereal Chem.*, 32, 499–506.

SPAETH, J. N. (1936). 'Catalase ratio' as an indicator of viability of tree seed. *Amer. J. Bot.*, 23, 691–2. (Abstract.)

SPENCER, G. E. L. (1931). Seed storage and germination. The use of cool storage in retaining the germinating power of some oily seeds. *Trop. Agriculture*, *Trin.*, 8, 333.

SPENCER, H. M. (1926). *Laboratory methods for maintaining constant humidity.* In *International Critical Tables*, 1, 67–68. McGraw-Hill Book Co., New York.

SREENIVASAN, A. (1939). Studies on quality in rice. IV: Storage changes in rice after harvest. *Indian J. Agric. Sci.*, 9(2), 208–22. (Abstr. in *Biol. Abstr.*, 14, No. 3406, 1940.)

STANSBURY, M. F. & GUTHRIE, J. D. (1947). Storage of cottonseed and peanuts under conditions which minimize changes in chemical composition. *J. Agric. Res.*, 75, 49–61.

STEINBAUER, C. E. & STEINBAUER, G. P. (1932). Effects of temperature and desiccation during storage on germination of seeds of the American elm (*Ulmus americana* L.). *Proc. Amer. Soc. Hort. Sci.*, 28(1931), 441–3.

STEINBAUER, G. P. (1937). Dormancy and germination of *Fraxinus* seeds. *Plant Physiol.*, 12, 813–24.

—— (1944). The embryo excision method of determining seed viability with particular reference to *Fraxinus* and *Sparganium. News Lett. Ass. Off. Seed Anal. N. Amer.*, 18(6), 4–6.

STEVENS, O. A. (1935). Germination studies on aged and injured seeds. *J. Agric. Res.*, 51, 1093–106.

—— (1943). Some 30-year tests on germination of alfalfa and clover seed. *Bi-m. Bull. N. Dak. Agric. Exp. Sta.*, 6(2), 8–9.

—— (1954). 40-year germination tests with alfalfa and clover. *Bi-m. Bull. N. Dak. Agric. Exp. Sta.*, 16, 111–12.

STEVENSON, F. J. & EDMUNDSON, W. C. (1950). Storage of potato seeds. *Amer. Potato J.*, 27, 408–11.

STEWART, F. C. (1936). The relation of age and viability to the popping of popcorn. *Bull. N.Y. St. Agric. Exp. Sta.*, 672, 7 pp.

STONE, A. L. (1938). Results of ten years of germination tests on the same samples of vegetable seeds. *Proc. Ass. Off. Seed Anal. N. Amer.*, 23/26(1930/33), 76–80.

STUBBE, H. (1935). Samenalter und Genmutabilität bei *Antirrhinum majus* L. (Nebst einigen Beobachtungen über den Zeitpunkt des Mutierens während der Entwicklung). *Biol. Zbl.*, 55, 209–15.

—— (1935a). Weitere Untersuchungen über Samenalter und Genmutibilität bei *Antirrhinum majus*. *Z. indukt. Abstamm.- u. VererbLehre*, 70, 533–7.

SUSKIN, M. (1953). The effect of display conditions upon the viability of packet seeds in California. *Bull. Calif. Dep. Agric.*, 42(3), 120–37.

—— (1953a). The effect of dry chemical fire extinguisher materials on the viability of seed. *Bull. Calif. Dep. Agric.*, 42(2), 63–64.

SVENSON, J. (1952). Neuere Erfahrungen mit der Schnellmethods zur Bestimmung der Keimfähigkeit des Getreides. *Brauwelt*, 1952(19), 421–2. (Abstr. in *Biol. Abstr.*, 26, No. 36032, 1952.)

SVIHLA, R. D. & OSTERMAN, E. (1943). Growth of orchid seeds after dehydration from the frozen state. *Science*, 98, 23–24.

SWAINE, G. (1954). Underground storage of maize in Tanganyika. *E. Afr. Agric. J.*, 20, 122–8.

—— (1957). Trials on the underground storage of maize of high moisture content in Tanganyika. *Bull. Ent. Res.*, 48, 397–406.

SWANSON, C. O. (1941). The effect of low temperature in preventing damage to wheat stored with high moisture content. *Cereal Chem.*, 18, 299–315.

TAKIGUCHI, Y. (1932). [On the Davis' modified catalase-method prophesying the germinating power of seeds.] *Bult. Sci. Fak. Terk. Kjušu*, 5(1), 103–16. (In Japanese, with English summary, pp. 115–16.)

TAMMES, T. (1900). Ueber den Einfluss der Sonnenstrahlen auf die Keimungsfaehigkeit von Samen. *Landw. J.*, 29, 467–82.

TANASHEV, G. A. (1938). Improvement on the indicator method for the rapid determination of the germinative power of cultivated plant seeds. *Chemisation Socialistic Agric.*, 7(4), 101–7. (Abstr. in *Chem. Abstr.*, 33, 8241, 1939.)

TASHIRO, S. A. (1913). A new method and apparatus for the estimation of exceedingly minute quantities of carbon dioxide. *Amer. J. Physiol.*, 32, 137–45.

TÄUFEL, K. & POHLOUDEK-FABINI, R. (1954). Relations between viability and free fatty acid content in rape and colza seeds. *Pharmazie*, 9, 511–14. (Abstr. in *Chem. Abstr.*, 49, 9745, 1955.)

—— —— (1955). Germinating power and citric acid content of stored plants. *Biochem. Z.*, 326, 317–21. (Abstr. in *Chem. Abstr.*, 50, 4304, 1956.)

TAYLOR, C. A. (1941). Germination behavior of tree seeds as observed in the regular handling of seed at the seed extractory and nursery, Norfolk, Nebraska. *U.S. Dep. Agric. For. Service. Prairie States Forestry Project.* pp.

—— (1948). Observations on the storage and germination characteristics of Angelica seed. *Proc. Amer. Soc. Hort. Sci.*, 52, 471–4.

TEJIMA, T. & MISONOO, G. (1938). [Ueber die Keimfähigkeit der unter niederer Temperatur aufbewahrten Samen.] *Proc. Crop Sci. Soc. Japan*, 10, 56–64. (In Japanese, with German abstr. in *Japan. J. Bot.*, 9(4), (174)–(175), 1939.)

TERÄSVUORI, K. (1930). Der Einfluss des Alters auf den Saatwert des Kleesamens. *Suom. Maataloust. Seur. Julk.*, 20(2), 81–108. (Abstr. in *Bot. Centr.*, 160(7–8), 200–1, 1931.)

TERVET, I. W. (1943). Improper storage of soya beans oftentimes causes injurious molds. *Seed World*, 54(10), 8–9, 46.

TEUNISSON, D. J. (1954). Influence of storage without aeration on the microbial populations of rough rice. *Cereal Chem.*, 31, 462–74.

THOMAS, A. S. (1946). *Cinchona* in Uganda. *Emp. J. Exp. Agric.*, 14, 75–84.

THOMAS, B. (1941). The importance of moisture content in the storage of grain. *Z. Ges. Getreidew.*, 28, 66. Also in *Wschr. Brau.*, 58, 157–8, 1941. (Abstr. in *Chem. Abstr.*, 37, 188–9, 1943.)

THRONEBERRY, G. O. & SMITH, F. G. (1953). The effect of triphenyltetrazolium chloride on oat embryo respiration. *Science*, 117, 13–15.

TIEGHEM, P. VAN & BONNIER, G. (1882). Recherches sur la vie latente des graines. *Bull. Soc. Bot. Fr.*, 29, 25–29.

TILLOTSON, C. R. (1921). Storage of coniferous seeds. *J. Agric. Res.*, 22, 479–510.

TODARO, F. (1898). Concerning the duration of the germinative ability of the seed of various cultivated plants. *Staz. Sper. Agr. Ital.*, 31, 525–63. (Abstr. in *Exp. Sta. Rec.*, 11, 157–8, 1899.)

TONIOLO, L. (1949). L'impiego dei telluriti nella determinaxione precoce della vitalita dei semi. *Ricerca Sci. e Riconstruz.*, 19(10), 1173–7. (In Italian only; abstr. in *Biol. Abstr.*, 24, No. 16764, 1950.)

—— & BENI, G. (1954). [The use of tetrazolium chloride for determining viability of cereal seeds.] *Ann. Sper. Agr.*, 8, 1975–88. (In Italian, with English summary. Abstr. in *Chem. Abstr.*, 49, 4091, 1955.)

TOOLE, E. H. (1939). Seed longevity and seed storage. *U.S. Dep. Agric. Bur. Pl. Ind.*, 5 pp. (Abstr. in *Exp. Sta. Rec.*, 82, 182, 1940.)

—— (1942). Storage of vegetable seeds. *Leafl. U.S. Dep. Agric.*, No. 220, 8 pp.

—— (1957). Report of committee on seed moisture content and seed storage. *Proc. Int. Seed Test. Ass.*, 22, 400–8.

—— & BROWN, E. (1946). Final results of the Duvel buried seed experiment. *J. Agric. Res.*, 72, 201–10.

—— & TOOLE, V. K. (1939). Germination of carpet grass seed. *J. Amer. Soc. Agron.*, 31, 566–7.

—— —— (1946). Relation of temperature and seed moisture to the viability of stored soybean seed. *Circ. U.S. Dep. Agric.*, No. 753, 9 pp.

—— —— (1953). Relation of storage conditions to germination and to abnormal seedlings of bean. *Proc. Int. Seed Test. Ass.*, 18, 123–9.

—— —— & BORTHWICK, H. A. (1957). Growth and production of snap beans stored under favorable and unfavorable conditions. *Proc. Int. Seed Test. Ass.*, 22, 418–23.

—— —— & GORMAN, E. A. (1948). Vegetable-seed storage as affected by temperature and relative humidity. *Tech. Bull. U.S. Dep. Agric.*, No. 972, 24 pp.

TOOLE, V. K. (1939). Notes on the viability of the impermeable seed of *Vicia villosa*, Hairy Vetch. *Proc. Ass. Off. Seed Anal. N. Amer.*, 31, 109–11.

—— & TOOLE, E. H. (1954). Seed dormancy in relation to seed longevity. *Proc. Int. Seed Test. Ass.*, 18(1953), 325–8.

TOUMEY, J. W. (1921). On the viability of tree seeds after storage for 10 years. *J. For.*, 19, 814.

—— (1930). Some notes on seed storage. *J. For.*, 28, 394–5.

TOWERS, B. & HARRISON, G. J. (1949). Viability tests of cotton seed. *Bull. Calif. Dep. Agric.*, 38, 25–31.

TRAUTWEIN, K. & WASSERMANN, J. (1929). Fortsetzung den Versuche über die Halbkorn-keimmethode. Schnell- und frühkeimprobe. *Z. Ges. Brauw.*, 52, 69. (Abstr. in *Centr. Bakteriol. Abt. 2*, 79(1/7), 139, 1929.)

TRECCANI, C. P. (1951). Germination tests for dormant seeds of fruit trees. *Ann. Sper. Agr.*, 5, 201–6. (Abstr. in *Brit. Abstr.*, B 111, 243, 1951.)

TSKOIDZE, V. (1936). [Ouskorenii metod opredelenia vskhojesti semian tungo.] [Méthode rapide pour la recherche du pouvoir germinatif des graines d'*Aleurites*.] *Bull. Acad. Sci. U.R.S.S. Sér. Biol.*, 1936(1), 143–50. (In Russian, French title and abstr. in *Bull. Soc. Bot. Fr.*, 85(5/6), 475, 1938.)

TUITE, J. F. & CHRISTENSEN, C. M. (1957). Grain storage studies. XXIII: Time of invasion of wheat seed by various species of *Aspergillus* responsible for deterioration of stored grain, and source of inoculum of these fungi. *Phytopathology*, 47, 265–8.

TUKEY, H. B. (1944). The excised-embryo method of testing the germinability of fruit seed with particular reference to peach seed. *Proc. Amer. Soc. Hort. Sci.*, 45, 211–19.

—— & BARRETT, M. S. (1936). Approximate germination test for non-after-ripened peach seed. *Plant Physiol.*, 11, 629–33.

TURESSON, G. (1922). Über den Zusammenhang zwischen Oxydationsenzymen und Keimfähigkeit in verschiedenen Samenarten. *Bot. Notiser*, 1922, 323–35.

TURNER, J. H. (1933). The viability of seeds. *Kew Bull.*, 1933(6), 257–69.

TURRILL, W. B. (1957). Germination of seeds: 5. The vitality and longevity of seeds. *Gdnrs.' Chron.*, 142(2), 37.

ULTÉE, A. J. (1933). Het bewaren van zaadkoffie [Storage of coffee seed]. *Arch. Koffiecult. Ned.-Ind.*, 7(2), 75–83. (In Dutch only; abstr. in *Biol. Abstr.*, 10, 4111, 1936.)

VAN TIEGHEM—*see* TIEGHEM, P. VAN.

VENKATRAM, C. S. (1950). Seed-borne fungi and loss of cotton-seed viability. *J. Madras Univ.*, 19(b), 79–112. (Abstr. in *Biol. Abstr.*, 26, No. 16170, 1952.)

VERHEY, C. (1957). Die Unbrauchbarkeit des Tetrazolium-Verfahrens zur Prüfung vom durch Trocknung verletzten Saatgut. *Proc. Int. Seed Test. Ass.*, 22, 321–9.

VERRET, J. A. (1928). Sugar cane seedlings. *Rep. Ass. Hawaii. Sug. Technol.*, 7, 15–23. (Abstr. in *Exp. Sta. Rec.*, 60, 334, 1929.)

VIADO, J. (1938). The cutting test as a practical method of testing viability of seeds. *Philipp. J. For.*, 1, 219–25. (Abstr. in *Biol. Abstr.*, 13, No. 1311, 1939.)

VIBAR, T. & RODRIGO, P. A. (1929). Storing farm crop seeds. *Philipp. Agric. Rev.*, 22, 135–46.

VINCENT, G. (1929). [Das Altern der Koniferensamen.] *Ann. Acad. Tchécosl. Agric.*, 4A (pt. 4), 453–87. (German summary, pp. 488–92; abstr. in *Ber. Wiss. Biol.*, 14(5/6), 301, 1930.)

—— (1937). Uskaldňování smrkových semen [Storage of spruce seed]. *Ann. Acad. Tchécosl. Agric.*, 12(4), 469–74. (In Czech, with German summary; abstr. in *Biol. Abstr.*, 12, No. 5408, 1938.)

—— (1937a). Uskladňování modrínových semen [Storage of larch seed]. *Ann. Acad. Tchécosl. Agric.*, 12(4), 557–60. (In Czech, with German summary; abstr in. *Biol. Abstr.*, 12, No. 5409, 1938.)

—— (1938). Die aufbewahrung der Fichten- und Lärchensamen [The storage of spruce and larch seed]. *Z. Forst- u. Jagdw.*, 70(1), 45–51. (Abstr. in *Proc. Int. Seed Test. Ass.*, 11, 158–9, 1939.)

—— & FREUDL, A. (1931). [Effects of early harvesting of coniferous cones on seed quality.] *Lesn. Prace*, 10(5/6), 248–56. (In Czech, with German summary; abstr. in *Biol. Abstr.*, 8, No. 16333, 1934.)

VON DEGEN—*see* DEGEN, A. VON.

WACH, A. (1942). Versuche zur Selenitfärbung des forstlichen Saatgutes. *Allg. Forst-u. Jagdztg.*, 118, 178–88, 210–18. (Abstr. in *Biol. Abstr.*, 17, No. 1559, 1943.)

WAGGONER, H. D. (1917). The viability of radish seeds (*Raphanus sativus* L.) as affected by high temperatures and water content. *Amer. J. Bot.*, 4, 299–313.

WAHLEN, F. T. (1929). Hardseededness and longevity in clover seeds. *Proc. Int. Seed Test. Ass.*, 9/10, 34–39.

WAKELEY, P. C. (1931). Some observations on southern pine seed. *J. For.*, 29, 1150–64.
—— (1931*a*). Successful storage of longleaf pine seed. *For. Wrkr. U.S.*, 7(1), 10.
—— (1945). Office report on an extended analysis of Barton's storage tests of southern pine seed. *U.S. Dep. Agric. For. Serv. Southern Forest Exp. Sta., New Orleans* (typewritten).

WALLEN, V. R., WALLACE, M. A. & BELL, W. (1955). Response of aged vegetable seed to seed treatment. *Plant Dis. Reptr.*, 39(2), 115–17.

WALLER, A. D. (1901). An attempt to estimate the vitality of seeds by an electrical method. *Proc. Roy. Soc.*, 68, 79–92.

[WALSTER, H. L.] (1943). How well does old seed germinate. (Abstract.) *Bi-m. Bull. N. Dak. Agric. Exp. Sta.*, 6(2), 10–11.

WARD, H. S. (1951). Effect of relative humidities during storage on available and reserve foods of blue lupine seeds. *Proc. Ass. Sth. Agric. Wkrs.*, 48, 138.

WARD, H. S., Jr. & BUTT, J. L. (1955). Hygroscopic equilibrium and viability of naturally and artificially dried seed of crimson clover, *Trifolium incarnatum*. *Agron. J.*, 47, 576–9.

WARK, D. C. (1942). The influence of storage in contact with certain seed-pickling dusts on the germination of grain. *J. Aust. Inst. Agric. Sci.*, 8, 22–25.

WEBSTER, C. C. (1948). The effect of seed treatments, nursery technique and storage methods on the germination of tung seed. *E. Afr. Agric. J.*, 14(1), 38–48.

WEIBULL, G. (1955). The cold storage of vegetable seed, further studies. *Rep. 14th Int. Hort. Congr.*, pp. 647–67.

WEISE, R. (1937). Bemerkungen zu Gurewitschs Methode, die Keimfähigkeit von Samen zu bestimmen. *Ber. Dtsch. Bot. Ges.*, 55, 338–40.

WEISENFELD, M. (1929). Die Vorzüge der Halbkorn-keimmethode. *Z. Ges. Brauw.*, 52, 95. (Abstr. in *Zbl. Bakt.*, 79(23/26), 505. Abt. II, 1929.)

WEISS, M. G. & WENTZ, J. B. (1937). Effect of *Luteus* genes on longevity of seed in maize. *J. Amer. Soc. Agron.*, 29, 63–75.

WELTON, F. A. (1921). Longevity of seeds. *Mon. Bull. Ohio Agric. Exp. Sta.*, 6, 18–25.

[WENT, F. W.] (1948). An experiment that may last 360 years. Project will give botanists new information about the heredity of plants and the life span of seeds. *Life Magazine*, 24(5), 57–58, 60.

WHARTON, M. J. (1955). The use of the tetrazolium test for determining the viability of grass seed. *Proc. Int. Seed Test. Ass.*, 20, 71–80.
—— (1955*a*). The use of the tetrazolium test for determining the viability of seeds of the genus *Brassica*. *Proc. Int. Seed Test. Ass.*, 20, 81–88.

WHEELER, W. A. & HILL, D. D. (1957). *Grassland seeds*. D. van Nostrand Co., Inc., Princeton, N.J., 734 pp.

WHITE, D. G. (1947). Longevity of bamboo seed under different storage conditions. *Trop. Agriculture, Trin.*, 24, 51–53.
—— & DELGADO, R. F. (1948). Seed storage. *Rep. P.R. Agric. Exp. Sta.*, 1947, 63–64.
—— & VILLAFANE, A. G. (1947). Seed storage. *Rep. P.R. Agric. Exp. Sta.*, 1946, 35.

WHITE, J. (1909). The ferments and latent life of resting seeds. *Proc. Roy. Soc.*, B. 81, 417–42.

WHITE, O. A. (1946). The germination of peas in Florida and King Tut's tomb. *Turtox News*, 24(1), 3 pp.

WHYMPER, R. & BRADLEY, A. (1934). Studies on the vitality of wheat. I: Prolongation of vitality of wheat seeds. *Cereal Chem.*, 11, 349–60.
—— —— (1934*a*). Studies on the vitality of wheat. II: Influence of moisture in wheat seeds upon imbibition and speed of germination. *Cereal Chem.*, 11, 546–50.
—— —— (1934*b*). Studies on the vitality of wheat. III: Vitality, and the action of heat on wheat seeds. *Cereal Chem.*, 11, 625–36.
—— —— (1947). A note on the viability of wheat seeds. *Cereal Chem.*, 24, 228–9.

WILLIAMS, M. (1938). The moisture content of grass seed in relation to drying and storing. *Welsh J. Agric.*, **14**, 213–32.

WILSON, A. K. (1953). Storage and germination of Douglas-fir seed in central Idaho. *Res. Notes, Intermountain Forest & Range Exp. Sta.*, *Ogden, Utah*, No. 5, 3 Mimeo pp.

WINTERS, H. F. & RODRIGUEZ-COLON, F. (1953). Storage of Mangosteen seed. *Proc. Amer. Soc. Hort. Sci.*, **61**, 304–6.

WOLLENWEBER, H. W. (1942). Über die Lebensdauer von Kartoffelsamen. *Angew. Bot.*, **24**(2), 259–60. (Abstr. in *Biol. Abstr.*, **23**, No. 16854, 1949.)

WRIGHT, E. C. BARTON-, BOOTH, R. G. & PRINGLE, W. J. S. (1944).Analysis of barley from King Tutankhamen's tomb. *Nature, Lond.*, **153**, 288.

WRIGHT, R. C. (1941). Investigations on the storage of nuts. *Tech. Bull. U.S. Dep. Agric.*, **770**, 35 pp.

WYTTENBACH, E. (1955). Der Einfluss verschiedener Lagerungsfaktoren auf die Haltbarkeit von Feldsämereien (Luzerne, Rotklee, und gemeinem Schotenklee) bei länger dauernder Aufbewahrung. *Landw. Jb. Schweiz*, **4**, 161–96.

YANAGISAWA, T. & ASAKAWA, S. (1953). Woody plant seed viability tests with tetra-zolium salts. *J. Jap. For. Soc.*, **35**(2), 40–42. (Abstr. in *Biol. Abstr.*, **27**, No. 23462, 1953.)

ZACHARIEW, B. J. (1939). Ein bequemes Schnellkeimverfahren zur Prüfung des Samens einiger Nadelholzarten. *Forstwiss. Zbl.*, **61**(8), 238–49.

ZELENY, L. & COLEMAN, D. A. (1938). The chemical determination of soundness in corn. *Tech. Bull. U.S. Dep. Agric.*, **644**, 23 pp.

CONVERSION FACTORS OF WEIGHTS AND MEASURES

Measures of Weight—Avoirdupois to Metric

1 dram (dr.)	27·343 grains	1·772 grams (grammes)
1 ounce (oz.)	16 drams	28·3 grams
1 pound (lb)	16 ounces	0·454 kilograms
1 stone (st.)	14 pounds	6·350 kilograms
1 quarter (qr.)	2 stones	12·701 kilograms
1 hundredweight (cwt.)	4 quarters	50·802 kilograms
1 (long) ton (2,240 lb)	20 hundredweight	1·016 tonnes

Metric to Avoirdupois

1 milligram (mg.)		0·015 grain
1 gram (gm.)		0·564 dram
1 kilogram (kg.)	1,000 grams	2·205 pounds
1 tonne (metric ton)	1,000 kilograms	0·984 ton

American Weights to Metric

1 pound		453·592 grams
1 cental	100 pounds	45·359 kilograms
1 (short) ton (2,000 lb)	20 centals	0·907 tonne

Metric to American Weights

100 kilograms (kg.)	1 quintal	2·205 centals
1 tonne	1,000 kilograms	1·102 (short) tons

Measures of Length—British to Metric

1 inch (in.)		25·400 millimetres
1 foot (ft.)	12 inches	30·480 centimetres
1 yard (yd.)	3 feet	0·914 metres
1 mile	1,760 yards	1·609 kilometres

Metric to British

1 micron (μ)	1/1,000 mm. (1/1,000,000 m.)	1/25,400 inch
1 millimetre (mm.)		0·039 inch
1 centimetre (cm.)	10 mm.	0·394 inch
1 decimetre (dm.)	10 cm.	3·937 inches
1 metre (m.)	10 dm.	1·094 yards
		3·281 feet
		39·370 inches
1 kilometre (km.)	1,000 m.	0·621 mile

Measures of Area (Based on 1 metre = 39·370 inches)
British to Metric

1 square inch (sq. in.)		6·452 sq. centimetres
1 square foot (sq. ft.)	144 sq. in.	0·093 sq. metre
1 square yard (sq. yd.)	9 sq. ft.	0·836 sq. metre
1 acre	10,000 sq.m.	0·405 hectare
1 square mile	640 acres	2·590 kilometres
		258·998 hectares

Metric to British

1 square millimetre (sq. mm.)		0·00155 sq. inch
1 square centimetre (sq. cm.)	100 sq. mm.	0·155 sq. inch
1 square metre (sq. m.)	100 sq. dm.	1·196 sq. yards
1 hectare (ha.)	10,000 sq. m.	2·471 acres
1 square kilometre (sq. km.)	100 ha.	0·386 sq. mile

Measures of Volume—British to Metric

1 cubic inch (cu. in.)		16.387 cu. centimetres
1 cubic foot (cu. ft.)	1,728 cu. in.	28·317 cu. decimetres
1 cubic yard (cu. yd.)	27 cu. ft.	0·765 cu. metre

Metric to British

1 cubic centimetre (cc. = ml.)		0·061 cu. inch
1 cubic decimetre (cu. dm.)	1,000 cu. cm.	0·035 cu. foot
1 cubic metre (cu. m.)	1,000 cu. dm.	1·308 cu. yards

Measures of Capacity —(based on 1 Imperial gallon = 4·546 litres).

1 pint (pt.)—0·568 litre	1 American gallon—3·785 litre
1 gallon (gal.)—4·546 litres	1 litre—1,000 cc.
2 pints—1 quart (qt.)	4 quarts—1 gallon

British to Metric

1 pint	0·568 litre
1 gallon	4·546 litres

Metric to British

1 millilitre (ml. = cc.)—0·061 cu. in.
1 centilitre (cl.)—10 ml.—0·610 cu. in.
1 litre (l.)—100 cl.—1·760 pints

Temperature

0° Centigrade (=Celsius)=32° Fahrenheit
The following formulae connect the two major thermometric scales :
Fahrenheit to Centigrade : $°C = 5/9 (°F - 32)$
Centigrade to Fahrenheit : $°F = 9/5 (°C + 32)$

SUBJECT INDEX *

Abelmoschus esculentus seeds, viability determined by tetrazolium salts and excised embryo technique compared, 128
Abies alba seeds, viability tested by use of indigo-carmine, 117
— *balsamea* seeds, comparison of germination, excised embryo, and tetrazolium tests, 130
— *concolor* seeds, viability test by excised embryo technique, 114
— *nobilis* seeds, longevity of, 79
Abutilon theophrasti seeds, life-span in soil, 9
Acer saccharum, leaf, double samara and seeds of, Fig. 1
— seeds, effect of desiccation on viability of, 94–95
— *tataricum* seeds, viability determined by tetrazolium salts and excised embryo technique compared, 128
Acorns, *see Quercus*
Acorus calamus seeds, longevity of, 92
Actinea odorata seeds, life-span in soil, 10
Agricultural seeds, viability tested with tetrazolium salts, 126–7
Agrostemma githago seeds, life-span in soil, 8
Agropyron seeds, effect of maturity on viability and longevity of, 38
Agrostis stolonifera seeds, longevity of, 68
Air, effect on longevity of *Gossypium* seeds, 34
— effect on longevity of *Oryza sativa* seeds, 35
— effect on longevity of *Populus* seeds, 35
— effect on longevity of *Zea mays* seeds, 34–35
Albizzia julibrissin seeds, life-span of, 3
Aleurites seeds, effect of desiccation on viability of, 98
— — viability tested by use of indigo-carmine, 117
Alfalfa, *see Medicago sativa*
Alisma plantago-aquatica seeds, longevity of, 91, 92
Alkaloid content, of deteriorating seeds, 152
Allium cepa seeds, absorption of moisture by, 15–21, 26
— — — effect of different packets on viability of, 101–6
— — — effect of moisture on viability of, 14–21
— — — effect of relative humidity on viability of, 18, 25–28
— — — effect of temperature on viability of, 18
— — — effect of temperature and humidity on storage of, 148

Allium cepa seeds, longevity in storage, 50–52
— — — safe moisture content for storage of, 21
— — — use of old, 141
Alopecurus aequalis seeds, life-span in soil, 10
— *pratensis* seeds, longevity of, 68
Alsine media seeds, life-span in soil, 8
Alternaria, longevity of, on seeds, 45–46
Amaranthus retroflexus seeds, life-span in soil, 7, 8
— — — respiration of, 12–13
— — — viability on moist glass-wool at 20°C., 12–13
— seeds, catalase activity as measure of viability of, 115
Ambrosia artemisiifolia seeds, life-span in soil, 9
— *elatior* seeds, life-span in soil, 7, 8
Amylase activity in rice seeds and acorns, 151
Andropogon seeds, longevity of, 69
Angelica archangelica seeds, longevity of, 76
Angiosperms, vii
Anthemis cotula seeds, life-span in soil, 8
Anthoxanthum odoratum seeds, longevity of, 68
Anthyllis vulneraria seeds, life-span of, 3
Antirrhinum majus seeds, effect of below-freezing temperature on, 33, 62
— — — longevity in storage, 61
— — — mutation rate affected by age of, 154, 156
Apium graveolens seeds, safe moisture content for storage of, 21
Apricot, *see Prunus*
Aquatic plants, dormancy of, 91
— — longevity of, 91–94
— — *see also* names of individual plants
Arachis hypogaea seeds, absorption of moisture by, 15–21
— — — changes in proteins during storage of, 150
— — — effect of age of seed on plant quality, 138–9
— — — effect of moisture on viability of, 14–21
— — — effect of relative humidity on viability of, 18
— — — effect of temperature on viability of, 18
— — — longevity of, 74–75
— — — safe moisture content for storage of, 21
Arnica montana seeds, longevity of, 76

* As in the general text of this book, no distinction is here made between seeds and seed-like fruits, all being commonly referred to as 'seeds'.—Ed.

199

AUTHOR INDEX

Abbott, E. V., 96
Åberg, E., 5
Acton, E. H., 150
Adams, J. D., 35
Afanasiev, M., 76
Ahlgren, G. H., 37
Akamine, E. K., 72, 89, 160
Aki, S., 54
Albo, G., 151
Alcock, A. W., 40, 67
Allen, G. S., 33
Altschul, A. M., 48, 172
Amato, F. D', 157
Andersen, A. L., 174
Anderson, J. A., 40, 67
Andrén, F., 47
Anthony, K. R. M., 73
Arai, M., 9
Armolik, N., 42, 43
Arndt, C. H., 74
Asakawa, S., 133
Atanasoff, D., 154
Atkins, W. R. G., 18
Avery, A. G., 153

Bailey, C. H., 64
Bailey, S. W., 49
Baird, P. D., 126
Baker, D., 152
Bakke, A. L., 67
Baldauski, F. A., 179
Baldwin, H. I., 79, 83, 114
Ball, G. J., 58
Balu, V., 161
Barner, H., 83
Barrett, M. S., 112, 113
Barrons, K. C., 54
Bartlett, L., 138
Barton, L. V., vi, 11, 14, 18, 21, 25, 28, 30,
    31, 32, 33, 35, 36, 50, 51, 55, 56, 58, 59, 61,
    62, 79, 80, 82, 84, 85, 87, 88, 90, 93, 96, 97,
    100, 101, 104, 106, 107, 110, 111, 112, 114,
    141, 142, 171
Barton-Wright—see Wright, E. C. Barton-
Basak, M., 176
Bates, C. G., 79
Battle, W. R., 37
Bayfield, E. G., 152, 172
Beal, W. J., 7, 8
Beattie, J. H., 50, 53, 74, 75, 138
Becquerel, P., 2, 35
Bejnar, W., 46, 75, 140
Belajeva, G. M., 186
Bell, W., 196
Beni, G., 124

Benne, E. J., 174
Bennett, N., 125
Bernal, D. M., 124
Bethmann, W., 119
Bhatia, D. S., 150
Bibby, K. M., 88
Binkley, A. M., 160
Black, M. A., 47
Blakeslee, A. F., 153, 171
Blum, P. H., 67
Blumer, J. C., 78
Blytt, A. G., 3
Bonnier, G., 6
Booth, R. G., 197
Borthwick, H. A., xi, 194
Boswell, V. R., 18, 53, 65
Bradley, A., 63, 64
Brand, C. J., 160
Brase, K. D., 88
Brett, C. G., 47
Brewer, H. E., 28, 70
Brison, F. R., 53, 89
Brocq-Rousseu—see Rousseu, N. Brocq-
Brodskis, B., 115
Brown, E., 9, 14, 46, 100, 101
Brown, E. O., 10
Brown, F. R., 53
Brown, J. C., 126
Brown, R., 3
Brücher, H., 120
Brunson, A. M., 66
Buglia, 149
Bulat, H., 124, 135
Bunker, H. J., 5
Burgerstein, A., 67
Burgess, J. L., 71
Burlison, W. L., 71
Burr, H. S., 136
Bussard, L., 67
Busse, I., 88, 95
Butt, J. L., 28, 70

Campbell, J. G., 152
Candolle, A. de, 1
Carruthers, W., 50, 63, 67
Carter, D. G., 160
Carter, E. P., 65
Cartledge, J. L., 153, 154, 155, 156
Cartter, J. L., 161
Champion, H. G., 79
Childs, J. F. L., 97
Chirkovskii, V. I., 75
Christensen, C. M., x, 40, 41, 43, 179, 185,
    187, 192
Christidis, B. G., 28, 73, 139